The Institute of Biology's
Studies in Biology no. 104

Biological Clocks

John Brady

M.A., Ph.D., D.Sc., M.I.Biol.
Reader in Zoology,
Imperial College of Science and Technology,
University of London

© John Brady 1979

First Published 1979 by Edward Arnold (Publishers) Ltd, London
First Published in the USA in 1979 by
University Park Press
233 East Redwood Street
Baltimore, Maryland 21202

Library of Congress Cataloging in Publication Data

Brady, John.
  Biological clocks.

(The Institute of Biology's studies in biology; no. 104)
  1.   Biological rhythms.   I.   Title.   II.   Series: Institute of
Biology. Studies in biology; no. 104.
OH527.B66   574.1   78-13779

ISBN 0-8391-0254-2

Printed in Great Britain

General Preface to the Series

Because it is no longer possible for one textbook to cover the whole field of biology while remaining sufficiently up to date the Institute of Biology has sponsored this series so that teachers and students can learn about significant developments. The enthusiastic acceptance of 'Studies in Biology' shows that the books are providing authoritative views of biological topics.

The features of the series include the attention given to methods, the selected list of books for further reading and, wherever possible, suggestions for practical work.

Readers' comments will be welcomed by the Education Officer of the Institute.

1978

Institute of Biology
41 Queen's Gate
London SW7 5HU

Preface

Plants and animals fit their different ecological niches not only by occupying different *parts* of the habitat, but also by occupying it at different *times*, dividing up the resources they compete for temporally as well as spatially. To do so they have evolved a variety of 'clocks' that timetable their activities round the year, across the day, or through the rise and fall of the tides. This book is about all these clocks.

The study of biological clocks in general, and of circadian (daily) clocks in particular, was scarcely considered scientifically respectable a few decades ago, but now is one of the active growth points in biology, cutting across ecology, ethology, physiology, biochemistry, biophysics and cybernetics. Recognition of the importance of biological timing for ecology, behaviour and physiology has become widespread, and its study therefore now features in many school and university courses. Specialist books exist on rhythms and on photoperiodism, as well as on other aspects of timing such as migration, but these are all too advanced or too lengthy for the beginner. No simple introductory text exists from which the student can get easily to grips with the basics of the subject. This book tries to fill that gap.

London, 1978

J. B.

Contents

1 Introduction

1.1 Early evolution of clocks

Life first evolved in a cyclical environment that fluctuated violently between the freezing darkness of the night and the searing radiation of the day. The ultraviolet (UV) component of this radiation was probably an important source of energy for the early organization of primitive bio-molecules, but with the same energy must also have destroyed them. Four thousand million years ago in the Precambrian when life was first evolving, the earth was not protected from the destructive 190–300 nm far-UV radiation as it is today by a layer of ozone in the upper atmosphere. The daytime far-UV irradiation was therefore many times more intense then than now, and early life would presumably not have survived for very long, had it not developed a temporal organization that restricted its more UV-sensitive processes to the night.

There is no fossil record of this history, of course, but simple protistans such as *Euglena*, and individual cells of multicellular plants and animals in tissue culture, both have daily physiological rhythms, so twenty-four-hour rhythmicity seems to be a fundamental property of eukaryotic (nucleate) life. It has, however, apparently been lost, or perhaps never been acquired by more primitive forms that lack nuclei (i.e. the prokaryotes: viruses, bacteria and blue-green algae).

1.2 Types of biological clock

These daily, or *circadian* rhythms, as they are more correctly called (p. 11), although primitive, ubiquitous and fundamental to life, are not the only form of biological timekeeping. There are at least four classes of clock-like biological process.

(1) A wide range of high frequency rhythmical processes such as heart beat, respiratory movements, and spike generation in repeater neurones (Fig. 1–1).

(2) Rhythms that are related to environmental cycles, particularly circadian (Fig. 1–2), tidal, lunar-monthly and annual rhythms (Chapter 2).

(3) The phenomenon of photoperiodism, i.e. the physiological adjustment to season regulated by clocks that measure the length of the day (Chapter 6).

(4) The interval-timer type of phenomenon that occurs when dormancy or diapause last for a fixed minimum duration (§ 6.5).

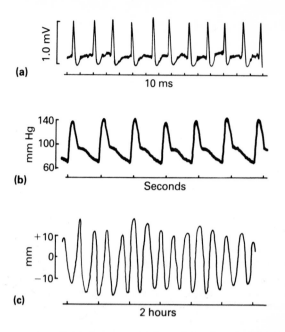

Fig. 1–1 Examples of high frequency biological rhythms. (a) Rhythm of action potentials in a nerve running to the eye muscle of the blowfly (*Calliphora*), sometimes called 'clock spikes' because of their great regularity (though see Fig. 1–3a). (Redrawn from BURTT, E. T. and PATTERSON, J. A. (1970). *Nature, Lond.,* **228,** 183.) (b) Human pulse, measured here as blood pressure in the femoral artery (cf. Fig. 1–3b). (Redrawn from BELL, G. H. *et al.* (1965). *Textbook of Physiology and Biochemistry,* 6th edition. Livingstone, Edinburgh and London, p. 460.) (c) Small oscillatory movements of the leaf of a bean seedling (*Phaseolus* sp.), measured as millimeters of rotation about the mean position (cf. Fig. 1–2a). (Redrawn from CUMMING, B. G. in 'PUDOC' (1972), after ALFORD, D. K., TIBBITTS, T. W. (1971). *Plant Physiol.,* **47,** 68.)

There is also a fifth type of physiological chronometry, the so-called 'continuously consulted clocks' which are involved in such processes as migratory navigation by birds or the 'alarm-clock' ability of some men (Chapter 5), but these involve continuously consulting circadian clocks (§ 5.1) and so do not rank as a separate class of clock-like process.

These four kinds of biological process are qualitatively different from other biological functions because they not only *consume* time, they also *measure* it, dividing it up into units. Strictly speaking, however, the first class of clocks are not true clocks, because they are not reset by external time-cues such as sunset or sunrise, and the speed at which they run is regulated, not against the passage of time, but by such factors as

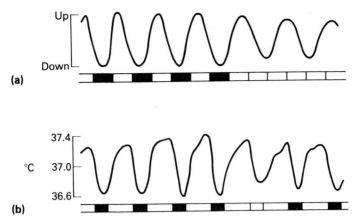

Fig. 1–2 Examples of 24-h, or circadian rhythms. (a) Plants of the bean family (Leguminosae) typically fold their leaves down at night and raise them up to the horizontal during the day; this figure is a kymograph trace of the movements of a French bean (*Phaseolus*) seedling (see Fig. 2–3 for method). Note: the rhythm continues for several days when the lights are left on (black bars on abscissa represent dark periods in a 12 h light : 12 h dark cycle). The high frequency rotatory movements of the leaf shown in Fig. 1–1c are much smaller than these daily movements and are not recorded here. (b) Daily rhythm of human rectal temperature (mean of 6 men living in an artificial light : dark cycle; black bars represent time asleep). Note: the night-time low still occurred on the night when the men stayed awake (see also Figs 4–2, 4–4). (Compiled from ASCHOFF, J. *et al.* (1972). *Aspects of Human Efficiency*, ed. W. P. Colquhoun, English Universities Press, London, p. 135).

temperature or physiological condition. Thus, the rate at which a man's heart beats depends upon his body's need for oxygen (Fig. 1–3b), and the rate at which rhythmic nerve spikes are delivered to the fly's eye is proportional to the temperature (Fig. 1–3a).

Because these high frequency rhythms work by oscillating, they meet, by definition, the criterion of measuring time, marking it out in units of the cycles of their rhythms. Moreover, they often do so very accurately, provided physiological conditions remain constant. But this does not make them true biological clocks, anymore than the clockwork motor of a toy car makes it a true mechanical clock.

The rest of this book thus deals with only those biological processes that (*a*) measure time by some means, (*b*) do so in relation to environmental time-cues (sunset, high tide, daylength, etc.), and (*c*) use this temporal information to control the timing of the organism's chemistry, physiology or behaviour: true biological clocks. The recently popularized and commercially exploited 'biological clock' that supposedly governs man's affairs over his lifetime – influencing his emotions, business success, driving, and so on – will not be considered, since no rigorous scientific evidence apparently exists to support it.

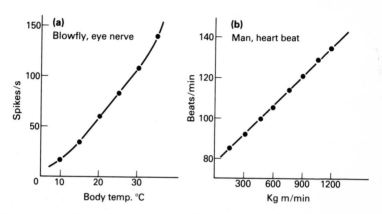

Fig. 1–3   The lack of constancy in high frequency rhythms. (a) The effect of temperature on the frequency of the nerve spike rhythm in the blowfly eye muscle shown in Fig. 1–1a. Between 15° and 30° the relationship is linear with a Q_{10} of just under 2 (see § 2.6). (Compiled from LEUTSCHER-HAZELHOFF, J. T. and KUIPER, J. W. (1966). *The Functional Organization of the Compound Eye*, ed. C. G. Bernhard. Pergamon, Oxford, p. 483.) (b) The relationship between the heart rate of a fit middle-aged man and his rate of energy expenditure (measured as Kg lifted 1 m per min) (cf. Fig. 1–1b). (After DE VRIES, H. A. (1967). *Physiology of Exercise*. Oxford University Press, London, p. 207.)

2 Daily, Tidal, Lunar and Annual Rhythms

Circadian (daily), tidal, lunar-monthly and annual rhythms in plants and animals are distinguishable from other biological oscillations by their temporal link with the environmental cycles from which they take their names. Since no other bio-rhythm has this kind of relationship with external time (see p. 2), this at once implies their adaptive significance: they permit the organism to adjust its life processes to the chronological arrangement of the external world. Thus, species divide up the environment between themselves not only on the basis of space, but also on the basis of time, so that each has its characteristic time of day (or tide, or year) for being active (Fig. 2–1). This requires (a) that the members of a species shall synchronize their activites, and (b) that each individual shall have a physiological timetable so that its phases of activity and rest are suitably prepared for metabolically. This timetabling is at the heart of all rhythms, be they circadian, tidal, lunar or annual, and is the reason for describing their control as being based upon biological clocks.

Both the observable rhythms themselves and their underlying clocks divide time into units by running repeatedly round some kind of stable physiological cycle, just as a pendulum does when swinging across one full traverse and back again. Also like a pendulum, these biological clocks therefore oscillate, passing repeatedly through the same physiological state at regular intervals. This close similarity with physical analogues is useful because it has allowed biology to borrow terms and concepts developed in the analysis of physical oscillators.

2.1 Terminology of biological rhythms

The basic concepts of an oscillator are simple. In the context of oscillating physiological cycles, the analogy of a rotating wheel is perhaps more appropriate than a swinging pendulum, though the principles involved are identical. If a point is marked on the tyre of a bicycle wheel, the bicycle is moved forward, and then the distance of the mark above the ground is plotted as a function of time, the result is a sinusoidal wave form of the type shown in Fig. 2–2.

One revolution of the wheel takes the mark from a to a' on the time base, and thus completes one cycle of the oscillation. The **amplitude** of the oscillation, which in this case equals the diameter of the wheel, is independent of time. The distance from a to a', however, is the *time* taken to complete one oscillation and is the **period** of that oscillation. As long as the bicycle moves forward at the same speed, the period of the wheel's oscillation will remain constant.

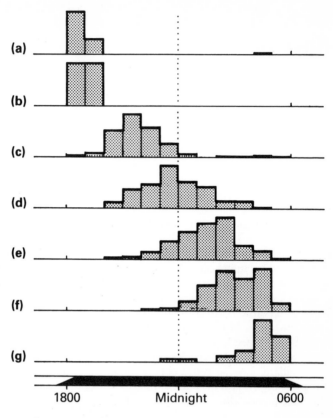

Fig. 2-1 The ecology of circadian rhythms. The difference in activity times of 7 species of African doryline ant. Height of columns shows relative numbers of flying males caught per hour of the night at light traps in Uganda. (a) *Dorylus moestus*; (b) *D. nigricans*; (c) *D. affinis*; (d) *D. fulvus*; (e) *D. alluaudi*; (f) *D. katanensis*; (g) *D. burmeisteri*. (Redrawn from HADDOW, A. J. *et al.* (1966). *Proc. R. ent. Soc., Lond. (A)*, **41**, 103.)

Another measure of the speed of the bicycle is the **frequency** of its wheel's rotation, i.e. the number of oscillations (revolutions) the wheel makes per unit time. This is the reciprocal of the period, so that as the bicycle accelerates, the period of its wheel's oscillation decreases and the frequency increases.

Any particular part or point in the cycle of an oscillation is a **phase**. Thus one refers to the phase of peak activity in an animal or the phase of minimal photosynthesis in a plant. If such a part of a rhythm occurs later (is 'delayed') or earlier (is 'advanced') in response to some environmental change, it is described as being **phase-shifted**.

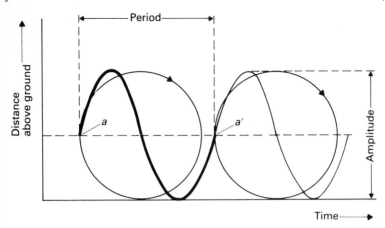

Fig. 2–2 The principles of oscillations and cycles. The thick line plots against time the sine wave displacement of point *a* on the wheel through one complete cycle of rotation. Note that the ordinate is true distance (amplitude of the oscillation) but that the abscissa is *time*, not distance. Because biological rhythms are generally not symmetrical about the mean line, biologists usually use the term 'amplitude' to identify the maximum, peak to trough deviation of the rhythm; the physicists' definition, however, is half that, i.e. the radius rather than the diameter of the wheel in this figure. (See § 2.1 for further explanation.)

Many biological clocks have rhythms whose oscillations approximate to **sine waves** (so called because the equation describing the curve in Fig. 2–2 is: $x = r \sin \omega t$, where x is the deviation from midline, r is the amplitude, ω the angular velocity of the wheel, and t is time). For example, the daily movement of leaves and the daily change in human body temperature both do so (Fig. 1–2). This has sometimes led to the assumption that a sine wave is a fundamental characteristic of those rhythms that approximate to it, but biological material rarely conforms to simple rules, and there seems to be no good reason why any biological oscillation should conform to the sine wave equation. A bean plant, for example, not only moves its leaves up and down roughly in a sine wave, but also moves them about when they are in the 'up' position (Fig. 1–1c). Similarly, there is often a distinct post-lunch dip in the afternoon rise of the human body temperature rhythm (whether or not a lunch has been eaten) (Fig. 4–2). The form of the wave is not particularly important, what matters is that most biological clocks work by oscillating, just as most man-made clocks do.

Note that it is usual to use the word **rhythm** to identify the overt, observable oscillations that the organism shows in its movement, metabolism, etc., and to restrict the word **oscillator** to the presumed, but unseen underlying clock that drives the overt rhythms.

2.2   Circadian rhythms

Many features of the physiology, growth and behaviour of animals and plants show daily, circadian rhythmicity. The easiest to measure are generally those involving mechanical displacements of some kind: leaf or flower movements in plants (Fig. 1–2a) and locomotor activity in animals. For plants that move their leaves, as do all members of the Leguminosae, the recording technique can be very simple (Fig. 2–3). For animals,

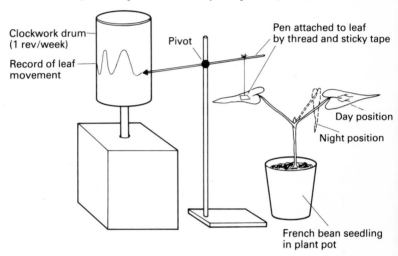

Fig. 2–3   Method for recording the leaf movement rhythms of plants – here a French bean seedling.

various refinements of the standard hamster-cage type of running-wheel have often been used (Fig 2–4). Running-wheels have the advantage over more sophisticated electronic recorders which respond to any displacement of the animal, in that they measure only walking or running, omitting from the record activities such as grooming or drinking that may be less precisely rhythmic.

Running-wheels are usually arranged so that rotation of the axle actuates a micro-switch connected to a pen-recorder. Each revolution of the wheel closes the switch and is thereby recorded by the pen as a lateral 'blip' on a line drawn on a continuously moving roll of paper. For experimental purposes a bank of several such wheels are set up in parallel, coupled to a multi-channel recorder.

Small mammals and large non-flying insects such as cockroaches have been the favourite subjects for running-wheel observations. Birds, which have also been extensively studied, have their activity recorded as 'perch-hopping', by the use of a spring-loaded perch that closes a micro-switch when the bird lands on it.

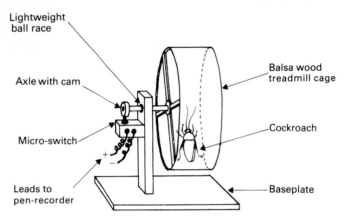

Lightweight ball race

Axle with cam

Micro-switch

Leads to
pen-recorder

Balsa wood
treadmill cage

Cockroach

Baseplate

Fig. 2–4 Running-wheel for recording the locomotor activity of large insects. As the cockroach walks or runs forward the wheel-cage rotates, and with each revolution actuates a micro-switch connected to a pen-recorder. A food pellet and water wick are arranged so that they do not touch the wheel but are continuously available to the cockroach through a hole in the centre of the right-hand face of the wheel.

All these kinds of activity recorders can be set up so that the animal (or plant) is provided continuously with food and water and can be left totally undisturbed in strictly controlled environmental conditions for many days or weeks. The activity record produced under these conditions then provides a measure of the animal's changing levels of spontaneous behaviour, i.e. its locomotor activity occurring in the absence of new stimuli from the environment. This is an important point in the study of rhythms, because it is necessary to know whether the measured rhythm is due to endogenous changes occurring within the organism, or due to its perceiving rhythmic changes in the environment. It is the latter situation, of course, that normally occurs in nature.

Let us analyse the activity of an animal treated experimentally in this way. A cockroach (*Periplaneta*) is placed in a running-wheel provided with food and water, and left undisturbed in a constant environment cabinet. After twenty days the paper is removed from the recorder and the trace displayed as shown in Fig. 2–5, each successive 24 h of record being arranged in chronological order down the page. For the first ten days the cockroach was kept in a 12 h light: 12 h dark regime (usually abbreviated **LD** 12 : 12; cf. **LL**, signifying constant light, and **DD**, signifying constant darkness) and it is clear that it performed the great majority of its running activity each day during the first two or three hours of darkness. There were sporadic bursts of activity at other times, but these were relatively rare – especially during the light phase – and reveal no obvious pattern. It is the daily synchronization of the peak of each day's activity

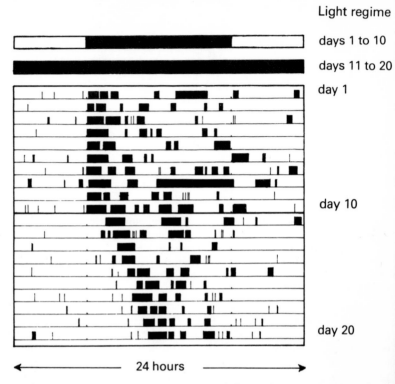

Light regime

days 1 to 10

days 11 to 20

day 1

day 10

day 20

← ——————— 24 hours ——————— →

Fig. 2–5 Activity record of a cockroach kept for 20 days in a running-wheel actograph of the type shown in Fig. 2–4. Each day's record presented in succession down the page. For the first 10 days the cockroach was in a 12 h light : 12 h dark cycle (LD 12 : 12) and then for the next 10 days in constant darkness (DD). Light regime indicated by bars at the top of the figure. For the significance of the free-running drift of the animal's activity rhythm during the 10 days in darkness see § 2.3.

to just after 'sunset' (i.e. the lights going out), forming a major black band descending straight down the record sheet, that identifies the strong daily rhythm.

Because of this particular animal started to run around almost as soon as the lights went out every evening, it appears that it may have been responding directly to the environmental time-cue of the artificial 'sunset'. That would imply that the behavioural rhythm was due merely to some direct photokinetic response to the lights going out – a rhythm driven directly by the environment and of no more interest than the one that could be demonstrated by, say, banging the side of the animal's cage once every 24 h to frighten it into activity.

However, for the last ten days of the experiment the lights were switched off permanently, and the cockroach left in constant darkness. What then happened to the activity pattern is highly revealing and crucial to the whole concept of circadian rhythms: the cockroach did not just run around sporadically, distributing its bursts of activity at random throughout the day, it maintained one major peak each 24 h, i.e. its daily rhythmicity persisted in the absence of environmental time-cues (see also Fig. 1–2).

Furthermore, examination of the record reveals that the activity peak did not occur at precisely the same time every day, as it had in the LD 12:12 regime, it now drifted relative to external, clock time, occurring about 0.5 h later each day. The cockroach thus showed a rhythm that both persisted in the absence of external time-cues, and did so at a period that differed from the 24 h by a small, but consistent amount each cycle.

2.3   Free-running rhythms

Released from the restraining influence of a cyclical environment, the cockroach's activity rhythm thus exhibited its own natural frequency and free ran, rather in the way that a badly adjusted mechanical clock will free run, fast or slow, unless it is corrected daily by its owner. This **free running** under constant conditions, at a period slightly different from 24 h, is typical – and diagnostic – of such rhythms, and is the origin of the term **circadian** used to describe them (from the Latin *circa diem*, about a day). This word avoids confusion with the old term **diurnal**, which strictly means during the daytime, and should be kept for use as the opposite of **nocturnal**, meaning during the night. The term **diel** is usually used to identify rhythms that have been observed under natural conditions or under artificial light : dark cycles, but which have not yet been shown to free run in constant conditions.

Circadian rhythms may diverge from the 24 h by a greater amount than the 0.5 h shown by the cockroach in Fig. 2–5, but they commonly fall within the range of 22–28 h per cycle. In artificial light cycles that do not add up to 24 h, they can be driven a few hours faster or slower than this (p. 17), but prolonged driving at unnatural frequencies can have pathological effects. It should also be noted that though many organisms show free-running circadian rhythms in constant light as well as in constant darkness, if the intensity is bright enough, constant light nearly always damps out any overt rhythmicity after a few cycles and stops the underlying clock (see p. 26).

Although free-running circadian rhythms drift relative to solar time and may diverge from 24 h by varying amounts, they may nevertheless run very accurately at their free-running frequency for many cycles, especially in homeothermic animals. The flying squirrel, *Glaucomys volans*, for example, can maintain its free-running period in constant darkness to an accuracy of ± 6 min per cycle for several weeks. Lower animals are generally less precise, however (cf. Fig. 2–5).

Free running is equally characteristic of circa-tidal, circa-lunar and circa-annual (circannual) rhythms (Fig. 2–6), and carries with it an important implication about the nature of all these rhythms. It is demonstrated experimentally in conditions that are effectively constant for the most obvious environmental variables: light, temperature, noise, mechanical disturbance, humidity, odour. Other variables, such as barometric pressure, magnetic field, cosmic radiation, and so on are usually not controlled, however. The question therefore arises as to whether the persistence of circadian (and the other) rhythms in normal experimental conditions is perhaps due to the organism responding to these other, uncontrolled variables. The answer appears to be, no. All these environmental factors vary with an exact 24-h cycle because they arise from the earth's daily rotation. Free-running circadian rhythms, on the other hand, run on cycles that are invariably slightly longer or slightly shorter than 24 h (the same kind of argument covers free-running tidal, monthly and annual rhythms). This strongly implies that the control of rhythms that persist in constant conditions originates within the organism, i.e. as an endogenous timing mechansim – a **physiological clock**. The argument of endogenous *versus* exogenous origins for rhythms is examined more fully in Chapter 3.

2.4   Tidal and lunar rhythms

As the moon passes over the surface of the earth its gravitational pull draws up a bulge in the seas lying underneath it and causes a reciprocal bulge on the opposite side of the earth. The movement of these two bulges of water therefore follows the moon round the earth causing high tides. Since the earth rotates on its axis relative to the moon once every 24.8 h, these high tides occur on average 12.4 h apart. When the moon is on the equator (as it is twice each month), the two tides are of equal height. At all other times, because of the geometry of the two bulges in the sea following the moon's declination, one tide is higher than the other.

The sun's gravity causes an additional complication by augmenting the moon's pull when the earth, moon and sun are all in line (as they are twice a month at the full and new moon), and by antagonizing its pull when the moon, earth and sun form a right-angled triangle with the earth at the 90° corner (as they do twice a month at the first and last quarters). In the first position the tides are greatest ('**spring** tides'), in the second position they are least ('**neap** tides'). There are therefore two sets of spring tides and two sets of neap tides each **synodic** (lunar) month of 29.5 days (the number of days between new moons).

The organisms that inhabit the inter-tidal zone are thus subject to both

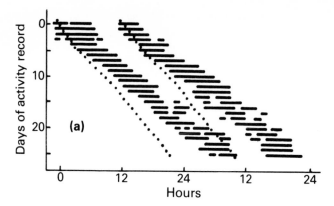

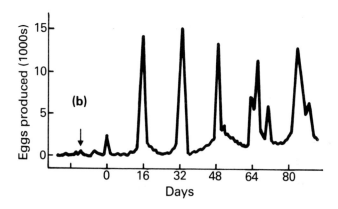

Fig. 2–6 (a) Tidal rhythm of locomotor activity in a fiddler crab, *Uca minax*, in constant (dim light) conditions in the laboratory. Each day's activity presented in succession down the page – as in Fig. 2–5. Dots indicate times of high tides on the crab's home beach. Note: (i) there are two unequal daily activity peaks corresponding to the two daily high tides, (ii) the rhythm shows a free-running period of more than 24.8 h and so drifts relative to real tide time. (Redrawn from PALMER (1974), after BARNWELL, F. H. (1966). *Biol. Bull.*, **130**, 1.) (b) Lunar monthly rhythm (actually half-monthly) of egg discharge in the marine brown alga, *Dictyota dichotoma*, in constant conditions in the laboratory but in a LD 14 : 10 cycle. At arrow the lights were left on for one night to simulate full moon; this was enough to initiate the rhythm. The abscissa scale is drawn in 16-day units from the first egg-discharge peak to emphasize that the rhythm is free running at about 16–17 days per cycle and so drifting relative to the real 14.8-day lunar cycle. (Redrawn from PALMER (1974), after MULLER, D. (1962). *Bot. Mar.*, **4**, 140.)

a 12.4-h and a 14.8-day cycle, alternating between being flooded with sea-water and running the risks of intense solar radiation, desiccation, freshwater irrigation, freezing, and predation by birds. Not surprisingly, all species have evolved strong tidal and semi-lunar rhythms by which they anticipate these hazards and take the necessary behavioural or physiological avoiding action.

Organisms that inhabit the middle tidal zone, such as the fiddler crab, are primarily subject to the 12.4-h cycle and show free-running rhythms of close to this frequency when studied in the laboratory (Fig. 2–6a). They often also show an endogenous 'awareness' of the unequal heights of the tide, spontaneously alternating their peaks of activity between a large one coinciding with the greater tide, and a small one coinciding with the lesser tide, as can be seen at the beginning of the record in Fig. 2–6a.

Organisms which inhabit the extreme upper littoral zone above mean high water are covered by the sea only at spring high tides; conversely, those inhabiting the extreme lower littoral, below mean low water, are *uncovered* only at spring low tides. These organisms are therefore primarily subject to the semi-lunar 14.8-day tidal cycle. A famous example is the grunion fish, *Leuresthes tenuis*, whose males and females swarm up Californian beaches to spawn on the crest of spring high tides. There they bury their eggs, leaving them to develop in the warm sand until the next high springs wash them out to hatch in the sea. Similar fortnightly rhythms occur in the activities of other denizens of these extreme tidal zones. The beach isopod crustacean, *Excirolana chiltoni*, for example, moults in a 15-day cycle.

Not many of these rhythms have been investigated in the laboratory, but the few that have exhibit spontaneous, free-running circa-semi-lunar rhythmicity. The brown alga, *Dictyota dichotoma*, for example, which has a 15-day rhythm of sperm and egg release in nature, maintains a *c.* 17-day rhythm of their release in tideless conditions in the laboratory (Fig. 2–6b). A particularly well-studied case is the midge, *Clunio marinus*, the larvae of which feed on red algae near the extreme lower limit of some European beaches. The adults live for only 2 h, and must therefore emerge, mate and oviposit during a single spring low water. When this midge is reared in constant conditions in the laboratory in LD 12 : 12, the adults emerge in a clear rhythm with peaks occurring at *c.* 15-day intervals. Both *Dictyota* and *Clunio*, and presumably most other similarly placed tidal organisms, therefore have an endogenous circa-semi-lunar clock controlling their activities.

2.5 Annual rhythms

There is increasing evidence that an appreciable proportion of long-lived animals exhibit circannual rhythms in the same sense that they exhibit circadian ones. When willow warblers (*Phylloscopus trochilus*), for example, are maintained in constant conditions and a never-changing LD 12 : 12 cycle for many months, they show what appears to be a free-running circannual rhythm of twice-yearly moulting: every *c.* 10 months they enter a 'spring' moult. A perhaps more striking example concerns the carpet beetle, *Anthrenus verbasci.* This takes 2–3 years to complete its larval development, and when cultures are maintained in constant darkness throughout this period, pupation does not occur at random, but in clearly defined peaks at about 42-week intervals, regardless of the external season.

It thus looks certain that some organisms possess circannual clocks that oscillate, in the absence of environmental time-cues, at a frequency of roughly once per year – strictly analogously to circadian clocks. The fact that in the few cases studied the period of the rhythm differs widely from the astronomical year seems to rule out the suspicion that the results are due to hidden environmental cues being picked up in the experimental 'constant' conditions. (The quite different phenomenon of seasonal photoperiodism is considered in Chapter 6.)

2.6   Temperature compensation

No clock is any use for measuring time if its speed of running varies with the temperature. All man-made clocks therefore incorporate mechanisms to compensate for temperature changes. So, too, do biological clocks. This, indeed, is one of the prime distinctions between a biological process that can be considered as a clock and one that should not (see pp. 2–3).

Birds and mammals presumably have little problem in this regard since they maintain a nearly constant body temperature (except in hibernation). Poikilothermic animals and plants, however, may experience body temperatures ranging through several tens of degrees centigrade. Since the rates of most physiological processes approximately double with each 10°C rise in temperature (i.e. like many chemical reactions they show $Q_{10} \sim 2.0$), this would create severe difficulties for their physiological clocks if they reacted to heat similarly.

For example, if the circadian rhythm of the cockroach in Fig. 2–5 were subject to normal Q_{10} laws and free ran at 24.5 h per cycle at 20°C, it would be expected to have a period of perhaps 12.3 h at 30°C because the clock controlling its locomotor activity rhythm would be running about twice as fast. Conversely, it would be expected to have a period of *c.* 49 h at 10°C. Clearly, that would annihilate any possibility of controlling the animal's behaviour on a 24-h time base.

The actual relationship with temperature for the cockroach is shown in Table 1a. The change in frequency over a 10°C temperature range is small, representing a Q_{10} for the underlying process of just under one. Similar compensation for temperature changes occur in all the circadian

Table 1   Examples of temperature compensation in free-running circadian rhythms. ((a) from BÜNNING, E. (1958). *Biol. Zbl.*, **77**, 141; (b) from LEINWEBER, F.-J. (1956). *Z. Bot.*, **44**, 337.)

(a) *The cockroach*, Periplaneta: activity rhythm			(b) *The French bean*, Phaseolus: leaf movement rhythm		
Temp. (°C)	Free-running period (h)	Effective Q_{10} for frequency	Temp. (°C)	Free running period (h)	Effective Q_{10} for frequency
19	24.4 ⎫		15	28.3 ⎫	1.01
23	24.5 ⎬ 0.94		20	28.0 ⎬	
29	25.8 ⎭		25	28.0 ⎭	1.00

rhythms that have been examined, whether in animals or plants. The French bean, for example, shows a Q_{10} of just over one, for the temperature range 15° to 25°C (Table 1b), representing a slight frequency increase in the underlying clock process.

The same kind of temperature compensation occurs in circa-tidal rhythms. Thus, the frequency of the free-running activity rhythm of the crab, *Carcinus maenas*, appears virtually unchanged between 15° and 25°C. Temperature compensation is also known in the circannual rhythm of one poikilotherm: the free-running life-span rhythm in colonies of the marine cnidarian coelenterate, *Campanularia flexuosa*, has a period of around 370–380 days in constant darkness at each of three constant temperatures: 10°, 17° and 24°C.

2.7   Entrainment

It is evident that in nature organisms can rarely be under the kind of constant conditions where their rhythms free run. If they were they would soon get out of phase with the environment. What normally happens is that the inherent frequency of the endogenous clock controlling the rhythm is reset by the environment, each cycle. For example, the cockroach in Fig. 2–5 free ran at a period of about 24.5 h in constant darkness, so that it lost 0.5 h per cycle relative to external solar time. When it was in the LD 12 : 12 regime, however, this drift was removed and its rhythm was reset the necessary half hour earlier each day by the arrival of 'sunset'.

This daily re-setting (**phase-shifting**) to maintain synchrony with the environment is called **entrainment**, and is analogous to the adjustment one makes every day to a wrist-watch if it is running a little fast or slow. Generally, there is one particular feature of the environment that is used as the time-cue (known as a **zeitgeber**, time giver). Most commonly it is either sunset or sunrise, occasionally both.

Most organisms use the day–night alternation as their prime zeitgeber, presumably because of its unique reliability by comparison with other daily environmental changes, such as those that occur in temperature, humidity or barometric pressure. In the absence of day–night cycles, however, several other features may act as zeitgebers. When held in constant light or darkness, many organisms will entrain to a 24-h temperature cycle, and social animals such as humans or bees are certainly entrained by social zeitgebers from their own species. Tidal animals entrain to the 12.4-h cycles of mechanical disturbance of the water.

There is an apparent paradox between the fact that circadian rhythms will entrain to temperature cycles, and the fact that they are compensated against changes in temperature. The point is that temperature compensation concerns only the underlying clock's ability to maintain stability when free running at different, relatively *constant* levels of ambient temperature, whereas temperature entrainment concerns the clock's ability to respond to a *rhythmic change* in the temperature by resetting its phase.

The fact that circadian and the other environmentally linked rhythms maintain rather regular periods in constant conditions, and do so over wide ranges of temperature, implies a marked stability in their underlying clocks. This stability is further demonstrated if attempts are made to drive them at frequencies widely divergent from normal by placing the organism in highly unnatural light cycles. Most circadian rhythms will entrain to light : dark cycles as short as about 18 h or as long as 30 h, which is a much wider range than they ever show spontaneously in constant conditions (see p. 11). Attempts to force them any further away from 24 h, however, results either in the rhythm breaking down, or in it locking on to alternate cycles of the zeitgeber, as may occur for example in a cycle of LD 6 : 6.

2.8   Phase response curves

An important aspect of entrainment is that organisms are not equally sensitive to zeitgeber time-cues all round the 24 h. The cockroach in Fig. 2–5, for example, could evidently make a 0.5 h phase-shift (i.e. re-set its rhythm) instantaneously when lights-out occurred only 0.5 h earlier than its endogenous clock would have made it active anyway (as shown by the difference between the activity times on days 10 and 11). But that does not mean that it could also have made the necessary immediate 5-h phase-shift to get it back into the correct, entrained phase if the original light cycle were switched back on at the end of day 20. Indeed, experiment shows that it almost certainly could not: the maximum phase-shift that most cockroaches will normally make to an earlier 'sunset' is about 3 h in any 24.

In common with all other organisms investigated in this way, the cockroach has a differential phase-shifting ability across the 24 h: at some points in its circadian cycle it can make bigger phase-shifts to a new zeitgeber time than it can at others. As Fig. 2–7 illustrates, when a new lights-out zeitgeber occurs 5 or 6 h before the activity peak is due, the cockroach advances its activity by about 2.5 h, but a few days later, when it is getting close to being fully re-entrained, and the zeitgeber now occurs only an hour before the expected activity time, the activity is advanced only about 20 min.

This differential phase-shifting to zeitgebers falling at different times in the rhythm's cycle is typical, indeed diagnostic, of all endogenous circadian and circa-tidal rhythms (it is presumably also typical of circa-lunar and circannual rhythms but has not been looked for in them). The effect has been much studied in circadian rhythms for the information it provides about the underlying clock.

The usual procedure is to place a series of individuals in constant darkness, let their rhythms free run, and then at different points (phases) in their free-running cycle expose each individual (or group of individuals) to a single 'pulse' of light of, say, 1 h duration. The effect this light pulse has on the timing of the free-running rhythm is then measured

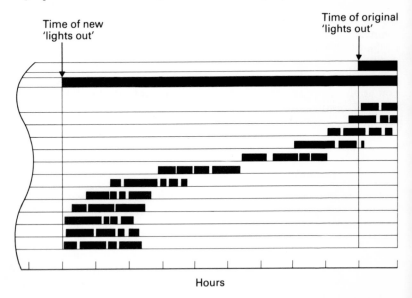

Fig. 2–7   Entrainment, in a cockroach. Activity record of an individual re-entraining to a new light cycle which occurs 9 h earlier than the one it had been experiencing previously (cf. Fig. 2–5). Light cycles shown by bars at top of figure. Light phase half of record, containing very little activity, omitted for clarity. Note: (i) phase-shifting is gradual, without jumps of more than 2–3 h; (ii) as the activity gets nearer the new lights-out signal the daily phase-shift gets successively less.

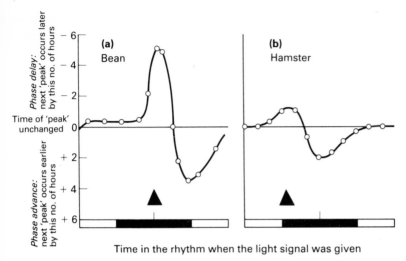

Fig. 2–8 Typical phase response curves: (a) for the circadian rhythm of leaf movement in the French bean; (b) for the circadian rhythm of running activity in the hamster. The timing of the bean's rhythm is identified by when the leaves are in their maximum folded-down night position, and the timing of the hamster's by when its activity peak starts. The phase-shifts of the rhythms in response to light signals can then be recognized and measured by changes in the time at which these 'peaks' (acting as reference phases) occur. The abscissa in each graph shows the 24 h of the rhythm's normal cycle, with a triangle marking the time of its 'peak' as it would occur relative to a 24-h light : dark cycle (white and black bars). Each graph then plots, on the ordinate, the amount by which the 'peak' is reset after the organism has experienced a single exposure to light (a light 'pulse') given (in otherwise constant darkness) at the times in its rhythm indicated along the abscissa. Each circle indicates the mean phase-shift of one batch of beans or hamsters given a light exposure at that specific time. The ordinate thus shows the number of hours by which the peak of the rhythm is observed to occur earlier (indicating that the rhythm has been phase-advanced) or later (indicating that it has been phase-delayed); and the abscissa shows the time in the c. 24 h of free-running rhythm when the light exposure was given to cause this phase-shift. In (a) the light signal consisted of a 3-h exposure to bright light in otherwise constant dim light; in (b) it was a 15-min light exposure in constant darkness. ((a) Compiled from MOSER, I. (1962). *Planta*, **58**, 199; (b) compiled from DAAN, S. and PITTENDRIGH, C. S. (1976). *J. comp. Physiol.*, **106**, 253.)

over the next few cycles of the rhythm with the organism still in constant darkness. One of three things happens: (1) the pulse has no effect and the rhythm continues to free run with no detectable change in phase; (2) the pulse falls at a time when it causes the rhythm to occur slightly earlier, thereby phase-advancing it (as occurred under different circumstances with the cockroach in Fig. 2–7); (3) the pulse falls at a time when it causes the rhythm to occur slightly later, thereby phase-delaying it.

After the rhythm has been tested by exposing a whole series of individuals in this way to single light pulses at different times right round the 24 h, it is then possible to plot a graph of the amount (and direction) of phase-shift against the time (phase) in the rhythm when the pulse was given (Fig. 2–8). Such a graph is called a **phase response curve** and is much used in rhythm research because it provides an identifying profile of the underlying, unseen oscillator that controls the overt, measured rhythm.

Phase response curves can be derived for any environmental signal that will act as a zeitgeber (light, temperature, mechanical disturbance (for tidal organisms), etc.). They can also be derived in many different ways. We have here considered only two: the kind just described (the commonest) and the kind derivable from experiments of the type shown in Fig. 2–7, where the new zeitgeber is repeated each cycle. It is worth pausing to ask, however, what adaptive significance phase response curves have. Because no organism is ever faced with these extreme phase-shifting conditions in the wild, one might expect that its rhythm would maintain a simple one-to-one relationship with new zeitgeber times (a 6-h zeitgeber advance causing a 6-h rhythm advance, and so on). The fact that none of the circadian and circa-tidal rhythms studied do have such simple phase response curves reflects the stability inherent in the underlying oscillator system (clock).

3 Endogenous or Exogenous?

There are in principle two ways that biological rhythms showing synchrony with environmental cycles could arise: (a) **exogenously**, as direct responses to some rhythmic change in the environment, or (b) **endogenously**, as responses to some internal physiological oscillator whose period is entrained to the period of the environment. It is the view of the great majority of workers in this field that circadian, circa-tidal, circalunar and circannual rhythms are all controlled by endogenous, physiological clocks, but it would only be honest to point out that that is not the universal view. There is a school of thought (principally associated with the name of F. A. Brown) that emphasizes the importance of exogenous factors in *causing* these rhythms, as opposed to just *entraining* them. The opposing sides of this controversy argue as follows.

3.1 Evidence for exogenous control

The case for the exogenous, environmental origin of rhythms is based on three main arguments. (1) The so-called constant conditons in which free-running rhythms are routinely demonstrated (see p. 12) are *not* constant with regard to many less obvious environmental factors; barometric pressure, magnetic field declination, gamma radiation, and other 'subtle geophysical synchronizers', as Brown calls them, all vary with a 24-h periodicity and penetrate standard laboratory 'constant conditions' so that they could in theory drive the observed rhythms. (2) Temperature compensation in circadian, tidal and annual rhythms is difficult to explain physiologically because the Q_{10}s for most physiological processes are around 2.0 (p. 15) but would require no explanation were the rhythm to originate directly from the environment. (3) It has proved impossible to find any actual biochemical clocks driving circadian or other rhythms, but such clocks would not in fact exist if these rhythms were all driven from outside the organism.

This third argument is a philosophical one that cannot be directly countered until a biochemical clock is unequivocally identified; the problem is considered from a physiological point of view in Chapter 7. The second argument is based in part on a misconception, as shown below (§ 3.2.2). This leaves just the first argument to be considered here. Accepting that there are several environmental variables that are not controlled in normal laboratory 'constant conditions', three questions arise: (1) are organisms capable of perceiving and responding to these unseen stimuli, (2) if they are, are their rhythms affected, and (3) if their rhythms are so affected, why do they free run in 'constant conditions' and drift relative to external solar time, even though the stimuli that are supposed to be causing them are all rhythmic with a precise 24-h periodicity because they arise from the earth's rotation about its axis?

To the first question the answer is, yes, there is much evidence that organisms can perceive these geophysical stimuli. Navigating birds respond to barometric pressure changes and to magnetic fields, bees respond to the daily cycle of magnetic declination, mice respond to changes in the background gamma radiation, and so on. Whether these responses mean that the answer to the second question is also yes, is less clear. Most of the evidence depends on complex correlations between changes in the candidate environmental factor and some aspect of the organism's rhythm. Since these correlations involve rearranging the data by sliding the time-scales, and correlation cannot in any case prove causation, the argument seems weak.

To the third question, of why organisms free run in 'constant conditions', the environmentalists' answer is the hypothesis of **autophasing**. A full explanation of this is given by PALMER (1974). In essence it runs as follows: (a) the observed rhythm is controlled and created by the organism's deep-seated response to the environment's geophysical 24-h (or lunar, etc.) oscillation; (b) as shown by its phase response curves (§ 2.8), this rhythm is differentially responsive round the 24 h to stimuli such as light and temperature; (c) in 'constant conditions' a certain level of temperature and light is ever present so that as the rhythm daily traverses its phase response curve for this level of light and temperature, it makes continuous adjustments by successive phase advances and phase delays; (d) the net outcome of these delays and advances is a small shift in the rhythm each cycle which thus appears as the so-called 'free-running' drift. The truth or falsehood of this hypothesis is difficult to prove without setting up an experiment in a bio-satellite out of earth orbit, but one point is clear now. Autophasing depends critically on the assumption that the rhythm is differentially sensitive to light and temperature across the 24 h, but there is no evidence for this under *constant* levels of light and temperature, only evidence for differential phase-shifting in response to *pulses* of light or changed temperature in otherwise constant conditions.

3.2   Evidence for endogenous control

The case for the internal, physiological control of rhythms (i.e. for endogenous biochemical driving oscillators) is based mainly on the following evidence.

3.2.1   Free-running drift

The fact that the rhythms of animals and plants held in constant light and temperature are maintained at periods which are consistently different from the external cycles of days, tides, months and years (§§ 2.3–2.5) is surely the strongest evidence for endogenous control. The autophasing rationalization of the free-running independence from the environment has to pre-suppose that the organism's rhythmic response to light and temperature is *qualitatively* different from its response to all other geophysical variables (and see also last para. § 3.1).

3.2.2   Temperature compensation

Biological rhythms are remarkably well temperature-compensated but never perfectly so, and the Q_{10}s for circadian rhythms usually fall within the range of *c*. 0.8 to 1.2 (§ 2.6). Were these rhythms to originate solely from the organism's response to external geophysical time-cues, no temperature compensation would be needed and their Q_{10}s would all be exactly 1.00. The fact that they are not, thus implies a physiological source of temporal homeostasis rather than control from the environment.

3.2.3   Time-zone translocations

If rhythms are due to geophysical time-cues they should adopt the timing of the local cues wherever the organism happens to be. However, when animals are maintained in constant conditions and flown rapidly across several degrees of longitude, they do not adopt local time unless they are exposed to local daylight. One of the best experiments was M. Renner's in 1957 on the circadian feeding rhythm of honeybees. He showed that bees trained in Paris to feed at 2000–2200 h local time and then flown overnight to New York, continued to feed as if on Paris time, i.e. 5 h too early for local, solar time in New York (Fig. 3–1a). The reverse experiment of training in New York and testing in Paris yielded exactly comparable results: on arrival in Paris the bees fed 5 h late, on New York time. Other translocation experiments confirm these findings: unless exposed to the local daylight cycle, organisms retain their original phase-setting when transferred to a new time-zone. One exception that is sometimes quoted is Brown's work on the shell-opening rhythm of oysters; close inspection of the data, however (see PALMER, BROWN and EDMUNDS, 1976, p. 216), suggests that the supposed shift to local time was no more than could be accounted for by the free-running drift of the rhythm in constant conditions. It may also be noted that any experiment which involves large phase-shifts (e.g. as a result of reversing the lighting regime) inevitably leaves the organism as markedly out of phase with local time as does a time-zone translocation, and there is likewise never any sign of phase-shifting back to local time.

3.2.4   Aperiodic rhythm initiation

The emergence of adult *Drosophila* cultured in constant light is completely arrhythmic; the clocks of all the individuals are evidently not working. If such a culture is transferred from contant light to constant darkness, however, the normal circadian emergence rhythm immediately appears (Fig. 4–5d): the first emergence peak occurs 15 h after the transfer to darkness and all subsequent peaks at *c.* 24-h intervals thereafter, notwithstanding that the transfer is a single, completely timeless, aperiodic signal. This seems easiest to understand on the basis that the clocks were stopped in constant light (as biological clocks usually are, see

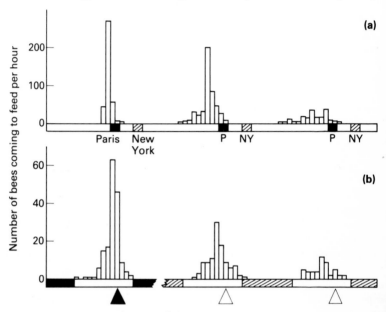

Fig. 3–1   Longitude displacement and time sense in bees. (a) Renner's Paris to New York displacement experiment. Bees were trained for several days to feed between 2000 and 2200 h local time in Paris in a controlled environment room equipped with a feeding point. They were then flown overnight to New York and tested there over the next 3 days for their feeding times in constant light in a similar room, but with no food provided at the feeding point. Abscissa: black blocks (at 24-h intervals) show time trained to feed in Paris; cross-hatched blocks show equivalent local time at New York (i.e. 5 h later). Note: the bees showed a free-running rhythm of visiting the feeding point with no sign of adopting New York time. (Compiled from VON FRISCH (1967), see Fig. 5–2, after RENNER, M. (1957). *Z. vergl. Physiol.*, **40**, 85.) (b) Comparable experiment showing bees' circadian time sense without displacement round the world. First histogram shows feeding pattern (in a LD 13 : 11 cycle) of bees trained to feed during the hour marked with the black triangle. Second two histograms show the feeding pattern (with no food provided) of the same bees over the third and fourth day in constant light (same time scale, with cross-hatching representing previous night phase, and open triangles previous training hour). Note: the feeding pattern shows a free-running rhythm, almost identical in form to Renner's in (a). (Compiled from BEIER, W. (1968). *Z. bienenforsch.*, **9**, 356.)

p. 11) and were simply made to start 'ticking' by the change to constant darkness. The fact that the rhythm is phase-locked to the time of the transfer to darkness, no matter at what local time that occurred, is further evidence of independence from geophysical time-cues.

3.2.5   Rhythms at the South Pole

The nearest approach yet made to studying rhythms out of earth orbit has been the work of K. C. Hamner at the South Pole in 1962. He set up experiments to measure the circadian rhythmicity of hamsters (running), *Drosophila* (emergence), beans (leaf movement) and the fungus *Neurospora* (mycelial growth) in constant conditions exactly on the pole but on a turntable that rotated counter-clockwise at one revolution per 24 h. Since the earth's 24-h geophysical cycles arise principally because of its daily rotation relative to the sun, this treatment should have removed virtually all geophysical 24-h periodicity from the experimental environment by maintaining the organisms in a constant geometrical relationship with the sun. Nevertheless, all four showed typical free-running circadian rhythms, regardless of whether they were rotated on the turntable *or not*.

To some extent all these pieces of evidence are circumstantial, or even negative, in the sense of being 'difficult to explain' on the exogenous basis. However, scientific inference is based on the relative plausibilities of the evidence for competing hypotheses, and the case for the endogenous origin of rhythms seems much more convincing than the opposing case for exogenous control. Certainly that is the verdict of the large majority of biologists.

4 Circadian Rhythms

For many organisms the most obvious manifestation of biological timekeeping is their circadian rhythmicity (§ 2.2). This influences almost every aspect of their life, making them hatch, grow, move, respire, photosynthesize, feed, digest, mate, sleep, etc., more at some times of day and less at others. Examples of these rhythms are discussed in this chapter, and how they are controlled round the 24 h in Chapter 7.

4.1 Rhythms in unicellular organisms

Prokaryotes appear to possess no kind of circadian rhythmicity, possibly because their life spans are frequently less than 24 h so that day-long periodicity may be largely irrelevant to them. The simplest eukaryotes, the protozoa and unicellular algae, however, do have circadian rhythms. *Euglena*, for example, is endogenously rhythmic in its cell division and rate of movement.

The most intensively studied unicellular organism is the luminescent, dinoflagellate alga, *Gonyaulax polyedra*. This possesses four clear rhythms: in photosynthesis, cell division, background luminescence, and flashing in response to mechanical disturbance (Fig. 4–1). These rhythms all free run in constant light of 1000 lux, disappear in constant light of 10 000 lux, and show all the typical characteristics of circadian rhythms: entrainment, phase response curves, and temperature compensation. In practice, they normally have to be measured in vials containing many thousands of cells, but can be shown to occur in each individual cell and to appear in the population because the individuals are in synchrony.

Much of general interest about biological clock organization has been revealed by work on *Gonyaulax* (mainly by B. M. Sweeney and J. W. Hastings). The four rhythms peak at different times during the 24 h (Fig. 4–1), so that each bears a different phase relationship to the cell's circadian cycle, a relationship, moreover, that is retained when the rhythms are free running in constant conditions at periods of up to 27 h per cycle. This strongly implies that the cell runs on a single circadian oscillator to which all four rhythms are coupled, especially since if one of them is chemically inhibited (e.g. photosynthesis by DCMU) and then allowed to recover, when it does so it returns immediately to its normal phase position in the cycle.

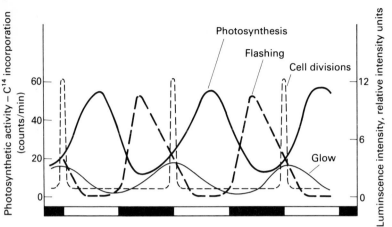

Fig. 4–1   Four rhythms in the unicellular alga *Gonyaulax* in a LD 12 : 12 cycle (abscissa). Photosynthesis measured as the rate of incorporation of radioactive $^{14}CO_2$ (left ordinate). Luminescence occurs as (i) 'flashing' in response to sudden mechanical disturbance, (ii) a background 'glow' (right ordinate, relative brightness units). Cell divisions occur almost exclusively around 'dawn'. Note: the 4 rhythms are differently phased relative to the day : night cycle. (Redrawn from HASTINGS, J. W. *et al.* (1961). *J. gen. Physiol.*, **45**, 69.)

4.2   Rhythms in cells

The fact that rhythms occur in Protista shows that circadian clocks can work in relatively simple systems and do not depend on feedback between cells; each protistan cell evidently contains its own clock. This also appears to be true for the cells of multicellular organisms.

Circadian rhythms have often been demonstrated in isolated bits of tissue cultured in constant conditions, for example in DNA and RNA synthesis in mouse liver slices, turgor pressure in bean leaf pulvini, and CO_2 production in leaf mesophyll fragments (Fig. 4–3a). Still more impressively, cell divisions occur at *c.* 24-h intervals in tissue cultures involving cell suspensions (e.g. in mouse fibroblasts). These observations all imply that the individual cells in the cultured tissues are being independently rhythmic, but the cultures are all multicellular systems (even in the suspensions the cells tend to be contiguous) so that the evidence for cell independence is not completely satisfying.

A rather clearer case concerns rhythms in the nuclei of cultured *Drosophila* salivary glands. L. Rensing demonstrated that the salivary glands of the larvae could be maintained *in vitro* for four or five days, and that it was possible with a phase-contrast microscope to measure the diameters of the nuclei in the undisturbed cells. When this was done at regular intervals round the day, the nuclei were found to swell and shrink in a clear daily pattern with two peaks per 24 h. Moreover, this pattern also occurred in constant light, so that the rhythm was endogenous and not simply driven by the culturing light cycle. Still more interestingly, however, when individual cells in a gland were watched continuously, their nuclear volume changes were found to get out of phase with their neighbours'. Hence, as with Protista, the rhythm was not dependent on some kind of synchronous chemical interaction between cells, but was a characteristic of each and every cell, individually.

This is currently the best demonstration of independent rhythmicity in the cells of a multicellular organism, but in view of what happens in tissue cultures and in Protista it seems very likely that all cells possess circadian clock mechanisms. Normally these cellular rhythms must cooperate in unison in the tissues they comprise, presumably by a combination of inter-cellular interactions and responses to rhythmic changes in hormonal levels (see pp. 57–8).

This unison provides the kind of tissue rhythms discussed below (§ 4.3), but some functions of cells pertain to themselves individually. The nuclear volume rhythms, for example, are simply a visible expression of complex cycles in nuclear activity, particularly in the synthesis of RNA and proteins. Some of these proteins will, as enzymes, cause substances to be released from the cell as secretions – to become in effect tissue secretions (enzymes, hormones, metabolites, etc.) – but most of the proteins are specifically concerned with the personal life of the cell, its energy production, maintenance, excretion, and so on.

4.3   Rhythms in tissues

Physiological rhythms are usually measured in whole organisms, but they commonly arise from the activities of specific tissues or organs. Excretory rhythms, for example, are due to the kidney, hormonal rhythms due to endocrine glands, and photosynthesis rhythms primarily to the leaf mesophyll. Other physiological rhythms, however, arise more generally in the organism, oxygen consumption and body temperature being obvious cases (Figs 4–3, 1–2b).

Perhaps the most important tissue rhythms in higher animals are those involved in the production of hormones. In vertebrates, the level of many hormones circulating in the bloodstream varies quite markedly round the day. The root of this lies in the hypothalamus, with its key position in the control of much of the body's homeostasis – especially as regards temperature, metabolic rate, and ionic balance. Through its control over the pituitary's production of adrenocorticotropic hormone (ACTH) it creates, or more probably entrains, the circadian activity of the adrenal cortex, and hence the rhythmic secretion of several corticosteroid hormones with wide-ranging and complex effects. The hypothalamus is thus an important driving component in the vertebrate's overall circadian rhythmicity. It is, moreover, closely linked anatomically with the supra-chiasmatic nuclei, which may be a master clock (see p. 50).

The adrenal corticosteroid rhythms were first identified by their daily variation in the urine, the changing make-up of which has provided a convenient measure of many physiological rhythms, especially in man. Such urinary rhythms arise in two ways: (a) from changes in the level of metabolites, hormones, etc., circulating in the blood plasma, and (b) from changes in the level of the kidney's excretory activity.

Sodium, potassium, chloride, and hydrogen ions all remain at fairly constant levels in the blood throughout the day but appear rhythmically in the urine because of the kidney's rhythmic treatment of them (Fig. 4–2).

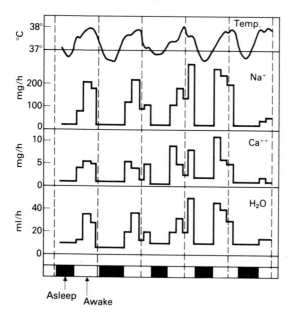

Fig. 4–2 Urinary rhythms in a human subject living without time-cues in isolation in an underground experimental 'bunker'; first 5 days' record of a 24-day sojourn. The K+ excretion rhythm (not shown) followed the Na+ rhythm very closely. Upper curve shows body temperature. Note the approximately synchronous drift of the four rhythms (five, including sleep) relative to solar time (dotted vertical lines at 24-h intervals). (Modified from ASCHOFF, J. *et al.* (1967). *Jap. J. Physiol.*, **17**, 450.)

This may be due to a rhythm inherent in the kidney itself, or be created by hormonal rhythms from the pituitary (anti-diuretic hormone) and adrenal (cortisol). Whichever is the case, the control primarily concerns the excretion of cations, since the removal of chloride is coupled with the removal of sodium, and the amount of water passed into the filtrate is proportional to the total amount of cations excreted. Other ions and metabolites, however (e.g. urea, calcium phosphate and the corticosteroid hormones), pass through the kidney in proportion to their concentration in the blood, so that their rhythmic appearance in the urine is a consequence of physiological activities elsewhere.

Circadian physiological rhythmicity is also expressed as daily changes in sensitivity to treatment with all sorts of chemical agents: drugs, toxins, allergens, antibiotics, hormones, anaesthetics, insecticides, etc. For example, bollworms (*Pectinophora gossypiella*) are 50% more resistant to organo-phosphate insecticides at dawn than at dusk; and mice show a $\pm 60\%$ daily variation in their sensitivity to ouabain (a specific inhibitor of Na^+/K^+ active transport). Although many such sensitivity cycles have been measured, the results generally reveal little about the physiological basis of circadian timing, because the pharmacological effects of the experimental treatments are far too widespread.

Less study has been made of tissue rhythms in plants. Root pressure, however, is one clear example of a tissue rhythm that has been shown to be circadian – in sunflowers. Another is the rhythmic fixation of CO_2 by succulent plants of the family Crassulaceae (e.g. *Bryophyllum* spp.). Their peak of CO_2-fixation occurs at night and hence is not photosynthetic; it arises, rather, from changes in organic acid metabolism involving re-fixing respiratory CO_2, a characteristic of the Crassulaceae associated with their water-conservation mechanisms. The interest in this rhythm centres on the biochemistry of its key enzyme (phosphoenolpyruvic (or PEP) carboxylase).

The CO_2-fixation rate varies across the 24 h from near zero around mid-day to a night-time peak of *c*. 200 μg CO_2/h/g fresh weight of leaf (the rhythm is maintained for a few days in constant darkness, and is the cause of the reciprocally timed rhythm in CO_2-output shown in Fig. 4–3a). Inspite of this great variation, the level of PEP carboxylase in the leaf tissues remains nearly constant – a relationship between an enzyme and its activity that also occurs in the photosynthetic rhythm in *Gonyaulax* (Fig. 4–1). It thus appears that biochemical rhythms of this type do not arise because the cell rhythmically synthesizes and breaks down the enzyme, but because the enzyme itself is rhythmic in its activity, perhaps because of the cyclical accumulation of an inhibitor. No equivalent enzymatic studies have yet been made of animal tissues, but this principle of cyclical enzyme activation, rather than cyclical DNA-transcription and enzyme synthesis, may prove to be an important feature of clock mechanisms (see § 7.4.1).

4.4   Rhythms in whole organisms

All rhythms arise principally from the activities of specific tissues (§ 4.3). Because of their widespread origin or effects, however, some are measurable only in whole organisms. Three important examples of this kind are metabolism, behaviour and development.

4.4.1 Metabolic rhythms

Metabolic rate is most easily measured in terms of O_2-consumption or CO_2-output. Changes in these arise from all the energy-demanding processes that vary in the organism, for example, locomotion, digestion, excretion, or growth. O_2-consumption rhythms therefore arise as an integration of all such activities. In plants they tend to be fairly simple and smooth-wave (Fig. 4–3a), but in active animals they are often more complex, with most of the pattern due to locomotor activity (Fig. 4–3b).

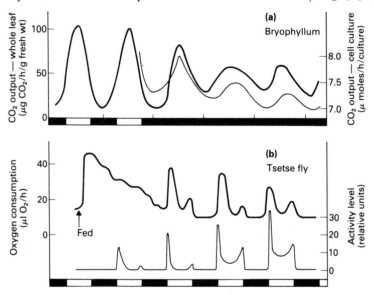

Fig. 4–3   Metabolic rhythms. (a) CO_2-output rhythm in isolated leaves (bold line, left ordinate) and in cultured mesophyll cells (thin line, right ordinate) of the succulent plant *Bryophyllum*; first in a light : dark cycle with a low light intensity (*c.* 500 lux) and then in constant darkness; abscissa shows light regime, with scale marks at 'midnights'. The rhythm arises indirectly from the reciprocal rhythm in dark fixation of CO_2 (see p. 30). (Modified from WILKINS, M. B. (1962). *Plant Physiol.*, **37**, 735; and (1965) *ibid.*, **40**, 907.) (b) Oxygen consumption rhythm (bold line) in a male tsetse fly (*Glossina*) which gorged with blood on day-1 (arrow); abscissa shows LD 12:12 cycle (modified from TAYLOR, P. (1977). *Physiol. Ent.*, **2**, 241). Lower curve shows typical activity patterns of male tsetse flies under the same light cycle (modified from BRADY, J. (1972). *J. Insect Physiol.*, **18**, 471; and (1975) *ibid.*, **21**, 807). Note: the energy demands of digesting the blood-meal initially obliterate any daily rhythm in oxygen consumption due to locomotor acitivity, and continue to cause a steady fall in oxygen consumption on days 3, 4 and 5 even though the level of locomotor activity is increasing (see § 4.4.1).

In mammals and birds, metabolic rate and motor activity are loosely coupled to the rhythm of body temperature (Fig. 4–4h); this is not necessarily a causal link, however, since the phase relationship may change when the rhythms are allowed to free run.

4.4.2 Behavioural rhythms

The most obvious circadian and tidal rhythms in animals are usually those involving behaviour. Spontaneous locomotor activity (i.e. running, swimming or flight in the absence of external stimulation) has been the most commonly studied aspect, mainly because of the ease with which it can be recorded. However, it probably also reflects a fundamental level of excitation in the central nervous system's overall control of behaviour, since other behavioural changes can be shown to run in parallel with it (Fig. 4–4).

Little direct research has been carried out on this, but three species have been investigated in some depth: man, the zebra finch (*Taeniopygia guttata*), and the tsetse fly (*Glossina morsitans*). Data on the tsetse fly cover the

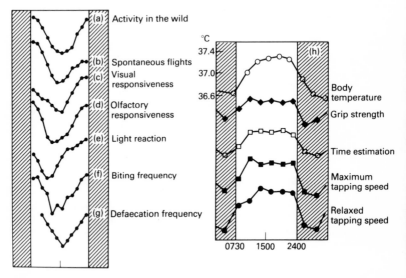

Fig. 4–4 Parallel rhythms in behavioural performance in an insect, the tsetse fly *Glossina morsitans* (a–g), and in man (h). Note that in each, quite different behaviour patterns, involving distinct neuro-muscular systems, change together across the day, implying that the central nervous system controls the thresholds of many responses synchronously on a circadian time-base. The measurements on the tsetse fly shown in curves b–g were made in a LD 12 : 12 cycle under constant conditions in the laboratory. (a–g after BRADY, J. (1975). *J. Ent. (A)*, 50, 79; h compiled from ASCHOFF, J. *et al.* (1972). *Aspects of Human Efficiency*, ed. W. P. Colquhoun, English Universities Press, London. p. 135.)

widest behavioural range. It is a markedly diurnal fly, being at its most aggressive in the morning and evening, largely inactive around noon, and almost totally so at night (Fig. 4–4a). This V-shaped pattern is circadian and persists in constant conditions in the laboratory (Figs 4–4b, 4–3b). The point of interest is that not only does the pattern of *spontaneous* flight follow the V-pattern, but so too does the pattern of *evoked* flight in response to experimentally applied stimuli (Fig. 4–4c, d), i.e. the fly's responsiveness to a periodically-presented standard stimulus varies across the day in a pattern indistinguishable from that which modulates its spontaneous activity. More interestingly still, behavioural responses that involve quite different central nervous and muscular systems also follow this V-pattern (Fig. 4–4f, g). The observations on rhythms in human performance (Fig. 4–4h) and in the zebra finch show the same effect.

It thus appears that in each species, many, widely differing behavioural responses change in parallel across the day. This implies that behaviour is organized with respect to circadian time by the central nervous system (CNS) going through a daily cycle of non-specific excitability. At one phase it is highly responsive to most stimuli (as in the tsetse fly at dawn and dusk); at another it is depressed and highly unresponsive to most stimuli (as in the tsetse fly at noon). It is therefore suggested that the part of the CNS that controls general arousal in relation to inputs such as hunger, danger or sex, is also coupled to a CNS clock that controls it on a circadian time-scale. Since responsiveness evidently fluctuates smoothly across the day (Fig. 4–4), this circadian control must modulate arousal continuously rather than in any sharp, on-off manner.

4.4.3   Developmental rhythms

Traumatic developmental events such as birth or hatching are typically rhythmic. Human births, for example, are 30% more common at night than they are during the day. Such rhythms are particularly striking in arthropods, which mitigate their acute vulnerability at moulting by doing so at times when the risks are least. This often results in the mass emergence of the adults of one species all at one time of day – as may be seen with midges on a summer evening.

The most extensive research has been done on *Drosophila melanogaster* and *pseudoobscura*, two species with a sharp circadian rhythm of adult emergence (eclosion), and which emerge mainly during the 6 h following dawn (Fig. 4–5b). Many years of work on this rhythm by C. S. Pittendrigh and his colleagues have revealed much of what is known about entrainment, phase response curves, temperature compensation, and the rhythm-stopping effects of constant light.

The adult eclosion of *Drosophila* appears as a rhythm only in a mixed-age population of flies. If all the larvae that are on the point of pupating are collected for one hour and then kept (at 20°C) separately from the rest, they do not later emerge in a rhythm, but as a single peak nine days later (Fig. 4–5c). The 'rhythm' is in fact a population effect; each individual fly, of course, emerges only once. Nevertheless, the eclosion is unquestionably under circadian control. If larvae are collected hourly as they pupate, throughout one day, the adults do not also emerge hourly throughout the day nine days later, but as three peaks, on the mornings of days 8, 9 and 10 (Fig. 4–5d).

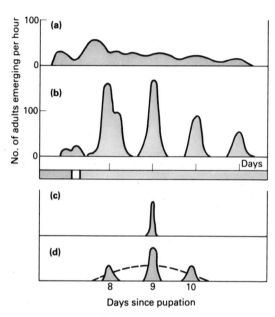

Fig. 4–5   Adult emergence rhythm of *Drosophila pseudoobscura*. (a) Distribution of adult emergence from a large, mixed-age culture maintained in constant darkness. (b) Adult emergence from a similar mixed-age culture maintained in constant darkness exactly as in (a) except that it received a single 4-h exposure to light shortly after the first adults appeared. Note that this converted the random distribution of emergence of (a) into a clear circadian rhythm. The abscissa shows the light regime, with 24-h intervals marked off from the first emergence peak. (Compiled from PITTENDRIGH, C. S. (1954). *Proc. natnl. Acad. Sci.*, **40**, 1018.) (c) The emergence of adults from a batch of *Drosophila* collected over a single hour as they pupated 9 days earlier in constant light; the batch was then transferred to constant darkness (see §4.4.3). (d) Emergence of adults from a batch collected as they pupated over a single 24-h period and then (as for (c)) transferred from constant light to constant darkness. Dashed line shows the distribution of emergences from another 24-h batch collected identically but then kept in constant light. Note that the transfer from constant light to constant darkness converts the random emergence in constant light (dashed line) into three 'gated' peaks on days 8, 9 and 10 after pupation (see § 4.4.3). (Compiled from PITTENDRIGH, C. S. (1966). *Z. Pflanzenphysiol.*, **54**, 275.)

Pittendrigh developed the concept of **'gating'** to account for this. The term implies that there is a gate 'open' for the flies to emerge through for only 6 h every day. Thus if their pre-adult development is complete before dawn on Monday they can emerge through Monday's gate, but if it is not complete until, say, noon, they cannot emerge then, and must wait for Tuesday's gate to open at the following dawn, and so on. Human births may be thought of as being gated in the same way (though less sharply); so too are a great many other kinds of such once-in-a-lifetime developmental events.

There are also developmental rhythms of a quite different kind, which are recognizably rhythmic in the individual. The growth rate of the oat seedling coleoptile, for example, varies with a circadian rhythm. And in a similar way the endocuticle of arthropods is secreted in a circadian rhythm of daily growth layers.

5 Celestial Navigation and Continuously Consulted Clocks

5.1 Continuously consulted clocks, or 'time sense'

One of G. Clauser's male subjects in 1954 could awaken himself intentionally after a night's sleep with an accuracy of ± 5 min. This particularly accurate example of a common experience appears impressive (a) because modern man is so dependent on his wrist-watch, and (b) because it occurs while he is asleep and unconscious. However, in view of the accuracy of the waking up of more primitive animals (§ 2.3), perhaps the effect is not so surprising.

Much more impressive is the ability of bees to feed with precise timing at *any* hour of the day, even when kept in constant, sunless conditions. This is important to them in nature because different species of flower open and secrete nectar at different hours, while they themselves spend most of their time in the darkness of the hive. The basis of this ability is their circadian rhythmicity: when trained to feed at a certain site and time in the laboratory, and then tested in constant light for the next few days, they visit the same, now foodless site at circadian intervals (Fig. 3–1b). It is in fact possible to train individual bees to feed at artificial food sources at up to nine distinct times in the sunless 12-h day of a laboratory. They cannot be trained to feed at times which do not make up intervals of 24 h, however, and will not, for example, feed at 19- or 27-h intervals.

Since they can do all this in constant conditions, bees evidently have the ability to 'read' the time of day from some internal clock, presumably their circadian one. Moreover, this 'reading' can be done at any time so the clock can be 'continuously consulted' – hence the term used to describe the ability.

At first glance, this seems to be a rather different phenomenon from the kind of circadian rhythms considered in Chapter 4, since those mostly concern the organism's relatively sudden change from one state to another, at circadian intervals. Many of those rhythms, however, are not on-off, binomial events, but smooth-wave quantitative changes across the 24 h (§ 4.4.2 and Figs 1–2, 4–4). They are therefore less like the periodic sudden flushing of a self-syphoning cistern (a relaxation oscillator) and more like the continuous swinging of a pendulum (a harmonic oscillator). The clock controlling them is not just a two-phase, high/low system, but fluctuates in 'level' continuously. Hence it is quite possible that bees, and other animals that need to consult their circadian clocks continuously (§ 5.2), do so simply by reading the phase that their circadian clock system is at, for example by monitoring their body temperature or state of central nervous arousal (p. 33).

5.2   Celestial navigation

Higher animals navigate around their environment using spatial or geographical cues. When distances greater than their immediate range of sensory perception are involved, they frequently use the sun as a principal cue, setting their direction at some specific angle to it. Commonly it is the sun's azimuth that is used, rather than any feature of its height (**azimuth** being the compass bearing to the point where a line drawn vertically down from the sun would meet the horizon; Fig. 5–3a).

However, since the sun moves across the sky by an average of 15° of azimuth per hour,* navigation by it without allowing for the passage of time must lead to serious errors in any journey lasting more than a few minutes. Some animals do indeed make such errors. The ant, *Lasius niger*, if shut away from the sun in the middle of a foraging trip, and then released a few hours later, sets off again at it original – but now erroneous – angle to the sun. Most animals, however, correct for the passage of time in this kind of experiment, and resume their journey at the correct compass direction, having changed their angle of orientation to the sun (Fig. 5–1a, b). In order to do this and to maintain a correct sun-based compass direction throughout a journey (i.e. navigate by 'sun-compass'), they must continuously up-date and change their sun-orientation angle. Like the bee seeking flowers at a specific time (§ 5.1), they must be able to 'consult' a clock at any time of day so as to predict where the sun should be.

Many species have been shown to possess this ability. The most closely studied are bees, because of the fascination of their dance language, and birds, because of their spectacular feats of global navigation. The juvenile, totally inexperienced shining cuckoo, *Chalcites lucidus*, navigates, alone, across 2500 miles of the Pacific Ocean from its foster parents' nest in New Zealand to its winter habitat in the Solomon Islands, presumably using a genetically-determined, celestial-based 'flight plan'.

Less impressive, but just as interesting from the clock point of view, are the sun-compass escape responses of some arthropods, among them the European sand hopper, *Talitrus saltator*. This, when dug out of its daytime burrow near the high tide mark, flees in a compass direction which would normally lead it down the beach to wet sand, but which is determined exclusively by the position of the sun and the time of day, and not by the configuration of the beach. An exactly similar navigational response, but in the opposite direction, occurs in the American frog, *Acris gryllus*, which swims back to the shore when thrown into deep water.

* The hourly change in azimuth actually varies markedly with season and latitude: at the equator, from 0°/h before and after noon to 180°/h at noon; at the poles, always *c.* 15°/h.

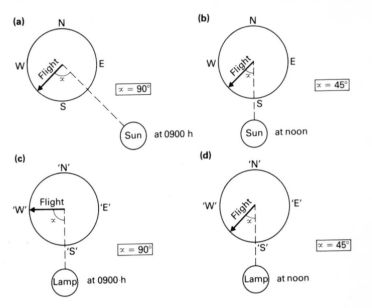

Fig. 5–1 Principles of 'sun compass' navigation in birds. (**a**) In order to fly SW at 0900 in the morning the bird must keep the sun at 90° to its left. (**b**) But by noon the sun has moved *c.* 45° clockwise round the horizon (i.e. in azimuth), see Fig. 5–2a) and the bird must therefore now keep it at 45° to its left in order to continue on the same SW course. That the bird knows how to adjust its flight angle relative to the sun by continuously 'consulting' an internal circadian clock can be shown in constant conditions in the laboratory (**c** and **d**). When the bird is kept in a circular arena with a bright lamp set to one side simulating the sun, it makes the same corrections in its directions of attempted flight relative to the lamp as it would relative to the sun were it outside in the open. Thus, as in (**a**) when attempting to fly SW at 0900 it keeps the lamp at 90° to its left (**c**), but 3 h later at noon it 'flies' with the lamp at 45° to its left (**d**), notwithstanding that it is kept in constant conditions and can receive no external time-cues. Furthermore, the bird makes the same directional changes if, instead of the lamp being moved, the bird's sense of time is changed by altering the timing of the light cycle it is in. Exactly the same principles of timekeeping are involved in all other animals that use a 'sun-compass' for navigation.

5.2.1   *Bird migration*

In birds, some of the best evidence for clock-corrected celestial navigation, i.e. for a sun compass, comes from work on starlings. These show migratory restlessness, known as **zugunruhe**, at certain photoperiodically-determined times of year, and G. Kramer showed in 1950 that when they are placed in a small circular cage that prevents them taking off, they make continual hopping movements mainly in one direction (which under an open sky would be their direction of migration). The compass bearing of these hops can then be recorded automatically and the bird kept in an artificial light cycle with a lamp to act as artificial sun.

When the 'sun' is kept still, the bird progressively changes its compass direction of zugunruhe across the day, after the manner shown in Fig. 5–1c and d. Conversely, if the 'sun' is moved, the bird makes the same kind of predictable changes in its hopping direction. It alters its angle of orientation systematically on the basis of clock time: at dawn the 'sun' is treated as if it were in the east, through the day the angle of orientation progressively moves round until at dusk the 'sun' is treated as if it were in the west. Many other experiments confirm these basic findings, and

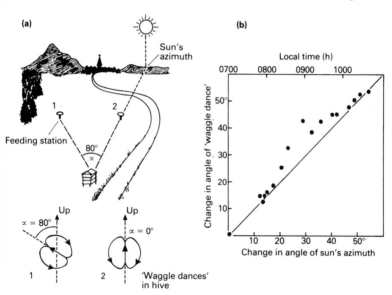

Fig. 5–2 'Sun compass' and time sense in bees. (a) Scene shows hive and two experimental feeding stations, No. 1 80° east of, and No. 2 in line with, the sun's azimuth. Lower figures show the form and orientation of the foragers' 'waggle dances' on the vertical comb in the darkness of the hive, indicating the flight directions to the two stations. Note: the axis of the waggle dance cross-over is vertically upwards ($a = 0°$) in the dance indicating station No. 2, but is at 80° to the left of the vertical ($a = 80°$) in the dance indicating station No. 1. (b) Angles of 17 waggle dances performed over a period of 3 h 40 min by a 'marathon dancer' indicating the direction of a new nest site, plotted against the change in angle of the sun's azimuth over the same period; diagonal line shows predicted, perfect relationship; the bee did not leave the hive during this period. (Modified and compiled from VON FRISCH, K. (1967). *The Dance Language and Orientation of Bees.* Staples Press, London.)

reveal in addition that birds also navigate by the moon and the stars, though in the latter case there is no need for clock correction since the pole star stays constantly due north (see HOFFMANN in 'PUDOC' (1972), and MATHEWS (1968) for more general reviews).

5.2.2   *Bee dancing*

The dance language of bees involves both celestial navigation and celestial communication. Foraging workers navigate to an established food source, such as a patch of flowers, by sun compass, and then communicate the necessary geometric information about the flight to potential recruits in the darkness of the hive by a figure-of-eight 'waggle dance'. In this, the angle of the cross-over of the eight relative to gravity on the vertical comb is the same as the indicated angle of flight from the hive relative to the sun (Fig. 5–2a), and is likewise based on a continuously consulted clock. As the sun moves round the day, the angle danced on the comb varies almost precisely as would be expected, even when the individual bee doing the dancing does not leave the darkness of the hive for several hours and so cannot see the sun (Fig. 5–2b). Sometimes this occurs right through the night and on into the next day, in which case the night-time dances point to the azimuth of the unseen sun on the other side of the earth.

6 Photoperiodism, or Seasonal Timing

6.1 Photoperiodism

Perhaps the best known aspect of biological timing is the fact that plants flower, and animals breed only at specific times of year. Early research indicated that flowering was determined by daylength; a suggestion first proved to be correct by W. W. Garner and H. A. Allard in 1920. This ability of an organism to measure the length of the day and set its reproductive or other developmental functions thereby is called **photoperiodism**. Flowering, seed germination and bud dormancy in plants; diapause and polymorphism in insects; moulting, nesting and migration in birds; mating and pelt growth in mammals; and much else besides all show photoperiodism (see LOFTS, 1970; VINCE-PRUE, 1975).

Photoperiodism, though specifically concerned with regulating the physiology of organisms round the year, is a quite different phenomenon from circannual rhythmicity (§ 2.5). A circannual rhythm is expressed in a roughly 50-week cycle when the organism is kept under either constant darkness or an invariant 24-h light:dark regime, no matter what the daylength. By contrast, photoperiodic induction of flowering, diapause, testis development, etc., can be invoked at *any time* by transferring the organism from a non-inductive light cycle to an inductive one – generally from short days to long days, or from long to short. By this means nurserymen make autumn plants such as chrysanthemums flower at Christmas or Easter, and chicken farmers make hens lay eggs all round the year. In nature this photoperiodic induction of development works together with any circannual rhythm the organism may have, but the two mechanisms are quite different, the one being determined by environmental light cycles, the other by an endogenous annual clock that can operate idependently of the environment.

There are two components to photoperiodism: an underlying, unseen biological clock that measures the day or night length, and an observable physiological reaction coupled to this clock. Thus in *Kalanchoe*, a succulent plant that flowers in the short days of spring, the clock inhibits flowering when daylength exceeds 12 h, but induces it when daylength falls below 11 h (Fig. 6–1a).

6.2 Critical daylength

Such developmental switching often occurs as the result of a small change in daylength, implying the existence of a critical number of

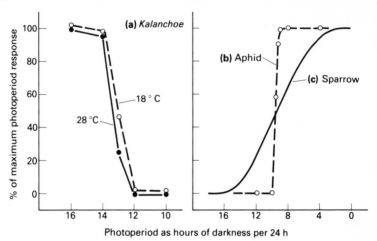

Photoperiod as hours of darkness per 24 h

Fig. 6–1 Demonstration of critical photoperiods in animals and plants, expressed as the percentage of the tested population showing the physiological response, plotted against the photoperiod (night length) they were exposed to. (a) Flowering induction in the succulent plant *Kalanchoe*. Note: (i) the sudden switch from zero flowering to near maximum flowering at around 13 h of darkness per day; (ii) that this critical night length is unaffected by the temperature at which the plants are grown, being indentical at 18° and 28°C. (After BÜNNING, E., 1967.) (b) Photoperiodic switching of female vetch aphids, *Megoura viciae*, from producing sexual offspring to producing parthenogenetic offspring. Note the sharp critical night length of 9.5 h. (From LEES, A. D. (1965). *Circadian Clocks*, ed. J. Aschoff, North-Holland, Amsterdam, p. 351.) (c) Testicular development in the white-crowned sparrow, *Zonotrichia leucophrys*. Note that the critical night length is relatively ill-defined. (From FARNER, D. S. (1965). *Ibid.*, p. 357.)

daylight hours for the induction to occur. This is termed the critical daylength (or more correctly the **critical photoperiod**, since either day or night length may be involved) below which the physiological response occurs in only a few individuals, but above which it occurs in most, or all of them. Often the critical photoperiod is very precise. In the vetch aphid, *Megoura viciae*, a 15-min increase in night length from 9 h switches some of the females from producing parthenogenetic offspring to producing sexually-reproducing offspring, and an increase to 10 h switches the entire population into doing so (with the 50% response at a night length of 9.5 h defining the critical photoperiod, Fig. 6–1b). In other cases, the switch is not so sharply defined, however (Fig. 6–1c).

Because the relationship between season and daylength varies with latitude, the critical photoperiod for a particular response in a given species changes systematically in populations from successively higher latitudes. Thus the moth, *Acronycta rumicis*, exists in several local races each with a different critical daylength for diapause induction: 14.5 h in the

Black Sea population (at 43° N), 16.5 h at Belgorod (50° N), 18 h at Vitebsk (55° N), and 19.5 h at Leningrad (60° N). The earlier onset of winter in the north is thus allowed for by the earlier onset of diapause while the days are still relatively long.

Critical daylength is often temperature compensated (§ 2.6), at least to some extent. Other aspects of photoperiodic responses are, however, greatly affected by temperature. For example, the critical daylength for flowering in *Kalanchoe* is 11 h at both 18° and 28°C (Fig. 6–1a), but whereas over 200 flowers are produced per plant in short days at 28°, only 6 are at 18°. Similarly, diapause in insects can be completely inhibited in photoperiods below the critical length, if the temperature is kept high enough. Thus the critical daylength for diapause in cabbage white caterpillars (*Pieris brassicae*) is between 15 and 16 h at all temperatures from 12 to 26°C, but above 26° few individuals do diapause. In such cases, therefore, although the photoperiodic clock is temperature compensated, the physiological processes it controls can be uncoupled from it at ecologically unsuitable temperatures. In other species, the relationship between temperature and photoperiod is still more complex, with the critical daylength being, in part, a function of temperature.

6.3   Photoperiodic clock mechanisms

Two quite different kinds of photoperiodic timing occur. (1) The organism apparently 'measures' the day or night length against its circadian periodicity, rather as if it were judging the time of sunrise and sunset by consulting its wrist-watch. (2) The organism has an interval timer which starts at sunrise and stops at sunset (or vice versa), quite independently of a circadian rhythm, rather as if it used an **hourglass** to measure critical daylength by turning it over at sunrise and waiting to see if the sun had set before or after the sand ran out.

In order to demonstrate these two mechanisms, somewhat complex lighting regimes known as 'skeleton photoperiods' are used. The organism is kept in a *day*length that is non-inductive for the process being examined, plus a very long night: e.g. cycles such as 12 h light: 60 h dark. Separate batches of the organism are then tested with a short light break at different times during the night, so that one batch may be kept under a cycle consisting of, say, 12 h light: 24 h dark: 1 h light: 35 h dark endlessly repeated, another batch under 12 h light: 36 h dark: 1 h light: 23 h dark, and so on. The start of the initial 12-h day acts as 'sunrise' and the end of the 1-h light break acts as 'sunset'. Thus, by experimentally scanning the whole long night length with light breaks in a series of batches of organisms, each batch being exposed to a cycle with the light break occurring at different times, it is possible to measure the physiological response (flowering, diapause, etc.) in a non-inductive daylength together with a synthetic 'skeleton photoperiod' that is independent of the normal 24-h cycle.

Two kinds of result occur. Organisms with circadian-based photoperiodic timing show physiological induction (or non-induction) at skeleton photoperiods that comprise multiples of c. 24 h (Fig. 6–2a). Those with 'hourglass-based' photoperiodic timing, however, show only a single peak of induction, at the one critical photoperiod (Fig. 6–2b; cf. Fig. 6–1b) with no sign of a circadian 'repeat' at longer skeleton

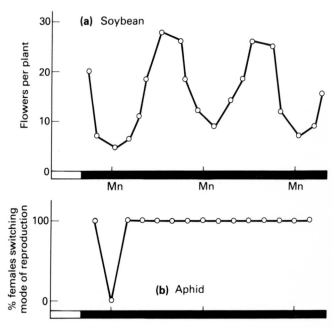

Fig. 6–2 The two types of photoperiodic response to 'skeleton' photoperiods. (a) The 'circadian' type, illustrated by flowering in the soy-bean, *Glycine max*, showing the mean number of flowers produced per plant in 19 cultures maintained in light cycles of LD 8:64 (ordinate) with 0.5-h light breaks at the times shown by the 19 circles. Note the 3 troughs in floral induction at circadian intervals. 'Mn' indicates times of what may be termed subjective midnights. (Redrawn from BÜNNING, E. (1969). *Photochem. Photobiol.*, **9**, 219.) (b) The 'hourglass' type, illustrated by the percentage of female *Megoura viciae* aphids switched from producing parthenogenetic offspring to producing sexually-reproducing offspring; 15 batches maintained in cycles of LD 8:64 with 1-h light breaks at the times shown by the 15 circles. Note that only the batch with the initial 8-h night failed to switch its mode of reproduction; no sign of circadian repeats occurred. In fact *Megoura* responds exclusively to critical night length, regardless of the preceding daylength. (Redrawn from LEES, A. D. (1966). *Nature, Lond.*, **210**, 986.)

photoperiods. From such experiments, reasonable evidence has been produced for circadian-based photoperiodic timing in flowering (e.g. *Kalanchoe*, soy-bean (*Glycine*), duckweed (*Lemna*)), in testis development in birds (quail, sparrows), and in diapause in insects (flesh-fly (*Sarcophaga*), parasitic wasp (*Nasonia*)). For hourglass timing, good evidence exists for sexual polymorphism in the aphid (*Megoura viciae*) and for larval diapause in the European corn borer moth (*Ostrinia nubilalis*).

Whichever kind of clock is used, several days and nights of the necessary duration usually have to be experienced by the organism before physiological induction occurs. The vetch aphid, for example, must be exposed to some six cycles with long days (in fact short nights are the critical feature, see Fig. 6–2b) before the parthenogenetic females will all switch from producing sexual offspring to producing parthenogenetic offspring. This implies that each inductive cycle by itself does not produce enough of some 'inductive substance' to throw the physiological switch and change the mode of reproduction, but that enough accumulates over several cycles to do so. Similar phenomena occur in the control of testis development in birds and flowering in some plants (e.g. *Kalanchoe* and *Glycine*), with the degree of the response increasing with the number of photoperiods experienced. Some species, however, need just one longer than critical photoperiod for induction to occur (e.g. flowering in *Xanthium* and *Lemna*).

6.4 Phytochrome

An essential component of any system that responds to light is a photo-sensitive pigment. Much effort has therefore been applied to the identification of the pigments involved in several different biological clocks, particularly in photoperiodic ones. The necessary experiments usually involve measuring 'action spectra', that is measuring photoperiodic responsiveness to light of different wavelengths to see which is the most effective and to identify the pigment by its absorption characteristics.

One system successfully explored in this way is the red-sensitive plant pigment, phytochrome (see KENDRICK and FRANKLAND, 1976). Phytochrome exists in two forms, red-absorbing (P_r) with peak absorption at 660 nm, and far-red-absorbing (P_{fr}) with peak absorption at $c.$ 730 nm. Exposure to daylight (i.e. white light that includes red) converts the P_r in plant tissues to P_{fr}. The reaction is spontaneously reversible, however, and in the absence of light settles at an equilibrium in which most of the phytochrome is in the P_r form. What happens under natural conditions, therefore, is that P_r is converted to P_{fr} during daylight, and reverts spontaneously to P_r during the night.

Phytochrome is involved, as both photo-receptor and activating agent, in several kinds of physiological switching in plants: e.g. seed germination, internode elongation, leaf movement, and photo-periodism. Because the dark reversion of P_{fr} to P_r takes an hour or two to complete, it was at first thought to act as a sort of photo-reversible hourglass for measuring night length. Later research has shown this to be unlikely, but it certainly is involved in other ways. It appears that P_{fr} either promotes or inhibts flowering depending on the phase relationship between its high concentration at the beginning of the night and the underlying circadian clock that measures the photoperiod length. In addition, it is implicated in the phase setting (i.e. entrainment, § 2.7) of that rhythm.

6.5 Dormancy and diapause

The ecological significance of photoperiodism is that it gears an organism's development to the seasons so that the organism is active at times when conditions are favourable (e.g. spring in temperate zones, rainy season in the tropics) and dormant when they are unfavourable (hibernating in winter, aestivating in the dry season). However, dormancy in plants, and the equivalent diapause in animals (especially in arthropods), once switched on photoperiodically, cannot be immediately broken by the return of favourable conditions: a minimum period in the dormant state must elapse first. This feature distinguishes dormancy and diapause from mere quiescence, which can be broken at any time by placing the organism in favourable conditions.

The relationship between water-availability, photoperiod, temper-ature and the duration of dormancy is complicated (see VINCE-PRUE, 1975; and LEES, 1955). The termination of dormancy is often accel-erated by some optimum temperature but delayed by either higher or lower tempartures. Thus, pupal diapause in the African moth, *Diparopsis castanea*, lasts 35 weeks at 22°C, drops to 25 weeks at its opti-mum 28°, but rises again to 45 weeks at 37°. In temperate species, such optimum temperatures usually bear some adaptive relation to the local mean winter temperature, commonly being a few degrees above freezing, e.g. 7°C for egg diapause termination in the silkmoth, *Bombyx mori*.

The duration of dormancy is therefore a quite different type of biological time-keeping from any that we have considered so far, and with quite different temperature compensation characteristics (cf. § 2.6). It appears to be a form of long-term interval timer, switched on by an appropriate photoperiod but then running for a fixed minimum number of days or weeks, regardless of external conditions. The mechanisms of these dormancy- and diapause-duration clocks are completely unknown, but presumably include the timed shutting down of endocrine systems that otherwise promote development.

7 Clock Mechanisms

7.1 Clock concepts

Any clock, whether man-made or biological, may be visualized as consisting of five components:

(1) *'mainspring'* – the source of energy that drives the whole mechanism;
(2) *'escapement'* (plus balance-wheel or pendulum) – the underlying, regulating oscillator;
(3) *'hands'* – the overt, measurable rhythms driven by the 'escapement';
(4) *'cogs'* – the coupling between the 'escapement' and the 'hands';
(5) *'adjuster'* – the mechanism for shifting the 'hands' to entrain them to external astronomical time.

Clearly it would be foolish to push the analogy with man-made clocks too far, but this conceptual breakdown does help to systematize analysis of the physiological mechanisms involved.

The 'hands', are all those overt rhythmic and other time-based phenomena that we have already considered. These are the only readily accessible component, and it is through them that the underlying processes of the clock mechanism have to be studied.

The 'mainspring', the metabolic energy needed to run the biological clock, is small, much less than is required by most other activities of the cell. Thus the phase of many circadian rhythms re-emerges unchanged when the animal or plant recovers from a period of anoxia deep enough to shut down all normally detectable metabolism. The spore discharge rhythm of the fungus *Oedogonium cardiacum*, for instance, is totally annihilated by treatment with NaCN but, as the cyanide concentration falls and its effects wear off, the rhythm re-emerges with its peaks exactly in phase with those of untreated controls. Some rhythms may be phase-shifted by several hours of deep narcosis, as is the emergence rhythm of *Drosophila* (which is delayed 10 h by keeping the pupae in pure N_2 for 15 h), but it may generally be concluded that, while dependent on aerobic processes, the energy demands of circadian clocks are very low. Nothing else is known of the 'mainspring' of biological clocks and the rest of this chapter is therefore devoted to the other three components: the 'adjuster', the 'cogs', and the 'escapement'.

7.2 Mechanisms for changing phase

Entrainment must involve three components: (1) a photoreceptor (or thermoreceptor, etc.) for detecting the occurrence of the zeitgeber;

(2) a means for coupling the receipt of this information to the driving oscillator; (3) a means for adjusting the phase of the oscillator.

Where the photoreceptive pigment participates directly in the chemistry of the clock, as it does in the photoperiodic hourglass of the aphid (§ 6.3) and may well do in all plants and Protista, these three components will be inseparable. In higher animals, however, the eyes usually act as the photoreceptor and, being physically separate from the clock, must be coupled to it by nerves. In the first case, therefore, light acts directly upon the clock to phase-shift it, and in the second it is electrical signals from the eyes that do so. Little is known of the mechanisms for changing phase except that they must conform to the phenomenology of the relevant phase response curves (§ 2.8).

7.3   Coupling processes

Apart from the obvious 'hands' themselves, the most successfully investigated of the clock's five components has been its 'cogs', the coupling between these 'hands' and the 'escapement' of the driving oscillator. Most of the work has been performed on arthropods, rats, hamsters and sparrows. The clearest story concerns the timing of emergence of silkmoths from the pupal case. Emergence does not occur the moment development is complete, but a few days later through a circadian 'gate' (see p. 35), and Truman and Riddiford demonstrated that the control of this timing is hormonal. The conclusions of their experiments, summarized in Fig. 7-1, run as follows: (a) emergence occurs normally at a species-specific time of day; (b) because emergence occurs, eventually, without the brain, emergence *behaviour* must be controlled by the thoracic and abdominal nervous system; (c_1) because re-implantation of the brain restores the rhythm, the source of emergence *timing*, i.e. the 'escapement', must be in the brain; (c_2) because the implanted brain has no nervous connection with the abdomen, its control of emergence must be exerted via a hormone that it secretes into the host's blood stream; (d) this hormone is not species-specific. These conclusions were reinforced by the demonstration that extracts of brain from late pupae cause almost immediate emergence, at any time of day, when injected into pupae containing fully-developed adults.

No other examples of gated rhythms have been investigated in this elegant way, but it seems quite likely that many such once-in-a-lifetime, gated events (§ 4.4.3) may be similarly coupled to the clock that controls them, by a pulse of hormone – at least in arthropods. It seems less likely that the same kind of push-button control can be involved in the daily-repeated, on-going rhythms that are typical of the rest of animal behaviour. Not only do these involve the continuous modulation of behavioural thresholds right round the 24 h (§ 4.4.2), but also the essential component of behaviour is that it shall always be readily

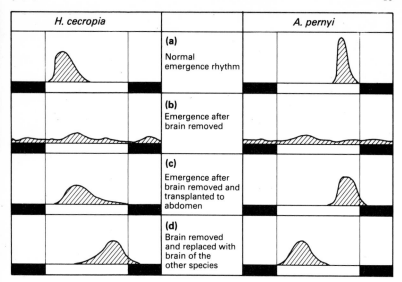

| H. cecropia | | | | A. pernyi |

(a) Normal emergence rhythm

(b) Emergence after brain removed

(c) Emergence after brain removed and transplanted to abdomen

(d) Brain removed and replaced with brain of the other species

Fig. 7–1 Summary of operations demonstrating the hormonal role of the brain in timing adult emergence of the silkmoths, *Hyalophora cecropia* and *Antherea pernyi*. Cross-hatching indicates the distribution of adult emergences resulting from the operations indicated. Emergence occurred arrhythmically when the brain was extirpated (**b**), but its circadian timing was restored by replacement of the brain into the abdomen (**c**); replacement with the brain of the other species resulted in the recipient pupae adopting the time of the donor brains (**d**). (Redrawn from TRUMAN, J. W. and RIDDIFORD, L. M. (1970). *Science*, **167**, 1624.)

adaptable in order to cope with problems as they occur. It is therefore not surprising to find no good evidence for the hormonal control of behavioural rhythms in any of the animals that have been thoroughly investigated, though hormones certainly play important roles in modulating behaviour in other ways.

The other insect circadian clock system that has been examined in depth is the one that controls the activity rhythms of cockroaches and crickets. It was once thought (and is still claimed in some textbooks) that this rhythm is controlled by a hormone secreted rhythmically by a small group of neurosecretory cells in the sub-oesophageal ganglion. This has since been shown many times to be most unlikely. It seems, rather, that the **optic lobes** of the brain are the crucial organs. The evidence, summarized in Fig. 7–2, implies that they contain the driving oscillators and that their relationship with the behavioural rhythm is quite different from that of the brain in silkmoth emergence. In cockroaches, the rhythm stops after operations **A** or **B** (Fig. 7–2), which leave the brain's neuro-endocrine system largely undisturbed, so that a hormonal coupling seems unlikely. On the other hand, these operations do interrupt all neural

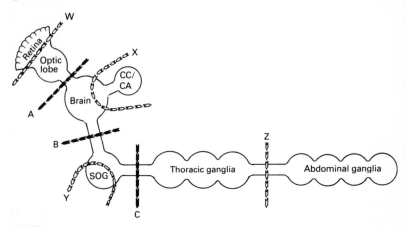

Fig. 7–2 Summary of operations on the cockroach to determine which part of its nervous system (shown here in side view) contains the clock driving the behavioural rhythm. When the main cephalic neuro-endocrine tissue (CC/CA) is removed together with the brain's neurosecretory cells (operation **X**), or when the neurosecretory cells of the sub-oesophageal ganglion (SOG) are destroyed (operation **Y**), the rhythm is unaffected. It is also unaffected if the abdominal ganglia are disconnected from the thoracic ganglia by severing the ventral nerve cord (operation **Z**). Cutting through the nerve cord at **B** or **C**, however, annihilates the rhythm and seriously disturbs the behaviour. Removing the retina (from both compound eyes; operation **W**) leaves the rhythm undisturbed except that it free runs in a light:dark cycle (the animal being effectively in constant darkness). Separating both optic lobes from the rest of the brain (operation **A**), on the other hand, does stop the rhythm, apparently without otherwise seriously disturbing the non-visual behaviour. It therefore appears that the optic lobes contain a vital component of the circadian system that controls behaviour. All the ganglia contain neurosecretory tissue, but none elicit rhythms when transplanted into other cockroaches. (After BRADY, J. (1969). *Nature, Lond.*, **223**, 781.)

connections between the putative clocks in the optic lobes and the effector organs (nerves and muscles) in the thorax, so that direct electrical coupling seems probable.

The circadian rhythms of mammals are probably entrained exclusively via their eyes, since blinded rats and mice show free-running activity rhythms when in light: dark cycles, even if the light is bright enough to penetrate deeply into their brains. The nervous route of this entrainment is not via the primary optic tracts, however, since these can be cut without upsetting entrainment. Instead, it is via a pair of fine nerve tracts that run directly from the eyes to two small groups of neurons that lie in the hypothalamus just above the optic chiasma (the **supra-chiasmatic nuclei**).

Micro-cautery destruction of the nerve cell bodies in the supra-chiasmatic nuclei annihilates all behavioural rhythmicity. Since this relatively minor operation leaves the neuro-endocrine systems virtually untouched, it appears that the nuclei may play the same primary driving role in mammalian circadian rhythms of behaviour that the optic lobes do in cockroaches and crickets, with the coupling from the clock likewise being neural. Destruction of the supra-chiasmatic nuclei also stops the circadian rhythms of melatonin secretion by the pineal organ and corticosterone secretion by the adrenal glands (see p. 29). From the point of view of understanding the overall circadian organization of animals (§ 7.5) the latter result is particularly interesting since hamster adrenal glands continue to show a clear rhythm of hormone secretion when isolated and cultured *in vitro*. Evidently there are different levels of rhythmic driving in the organism.

Sleep is objectively definable (by changes in the electro-encephalogram) only in vertebrates, and the sleep/wakefulness cycle is the most obvious manifestation of their circadian rhythmicity, with far-reaching physiological effects. It might be thought, therefore, that the control of sleep would be integral to the overall timing of vetebrate rhythms, but that appears not to be the case. In mammals, sleep control involves the release, within the brain, of the neuro-transmitter substance serotonin (5-hydroxy tryptamine) by neurons of the raphe nuclei in the brain stem, and the antagonistic secretion of the arousing neuro-transmitter adrenalin by neurons in the mid-brain. Some of the raphe neurons send axons to the supra-chiasmatic nuclei to which they supply serotonin, but in so doing they do not act in a primary driving role, since destruction of the raphe nuclei, while decreasing sleep in rats, does not stop their circadian rhythm of locomotor activity. The control of sleep in vertebrates thus appears not to be directly part of the system controlling their circadian rhythmicity.

In birds, the pineal organ plays a more important role than it does in mammals. Its removal depresses the amplitude of the perch-hopping activity rhythm, but does not necessarily stop it. There must therefore be a driving oscillator elsewhere in the CNS. In some species, however, this rhythm disappears a few days after pineal removal, and can be immediately reinstated by the implantation of a new pineal (into the eye, under the cornea). This implies that the system in birds is similar to that in mammals and insects, with the circadian control of behaviour effected by neural coupling from a CNS clock, but that it is different in the sense that hormones (from the pineal) may play a major part in determing how strongly the behaviour is expressed rhythmically.

7.4   Driving oscillator mechanisms

Whereas a certain amount is known about the coupling mechanisms between the primary driving oscillators and the overt rhythms they control (§ 7.3), almost nothing is known of how the driving oscillators themselves work. Four main theories exist: (1) that there is no endogenous driving oscillator, and that all rhythms are direct responses to the periodicity of the environment; (2) that slow, cyclical transcription of DNA occurs in the nucleus of the cell; (3) that frequency reduction occurs in relatively high frequency biochemical oscillations; (4) that slow, cyclical permeability changes occur in the cell's membranes. The first of these theories has been considered in Chapter 3 and shown to be improbable; the other three are discussed below.

7.4.1   DNA-replication and the 'chronon' theory

By the early 1960s evidence was accumulating that agents which disrupt nucleic acid chemistry have marked effects on the circadian rhythms of unicellular organisms. First, irradiation with ultraviolet at wavelengths known to be absorbed by DNA caused phase-shifts in the mating rhythm of *Paramecium*. Then, certain metabolic inhibitors were found to upset rhythms; in particular, antinomycin D (which blocks the transcription of DNA onto messenger-RNA) stopped the luminescence rhythm of the dinoflagellate *Gonyaulax* (§ 4.1) and the photosynthesis rhythm of the unicellular alga *Acetabularia*. This kind of evidence led to the 'chronon' concept of C. F. Ehret and E. Trucco. In outline, this proposes that transcription of a very long poly-cistronic DNA strand, the chronon, takes *c.* 24 h from end to end and then recycles. Each cistron along the chronon encodes sequentially for enzymes controlling processes that occur in the cell in a similar sequence round the 24 h, and the last cistron encodes for an 'initiator substance' which, when it diffuses back from the ribosomes to the nucleus, starts off transcription of the chronon all over again.

This is a satisfying hypothesis because it explains at a stroke why circadian rhythmicity is restricted to eukaryotes, where the genetic control of rhythms originates, and how the overall biochemistry of the cell is organized round the 24 h. It by no means explains everything, however. It indicates no clear mechanism for temperature compensation, nor does it explain how entrainment or phase-shifting occur. Perhaps most awkwardly of all, the chronon seems incapable of ever being in a state of arrhythmicity without stopping the overall running of the cell, and yet cells remain perfectly viable and biochemically active when they are running arrhythmically because they have been kept in constant bright light. On the other hand, some other kind of nuclear component in the clock mechanism is certainly not ruled out, especially since recent work suggests that inhibition of protein synthesis (e.g. by cycloheximide or anisomycin) can cause frequency changes in circadian rhythms.

7.4.2   Frequency reduction in high frequency biochemical oscillations

The fundamental problem in the construction of biochemical hypotheses for circadian (or tidal, etc.) oscillators is that all the known biochemical oscillations are temperature-dependent reactions with very short cycle times, at best lasting a few minutes (Fig. 7–3). There is a time gap of two to three orders of magnitude between such biochemical 'clocks' and circadian ones. It was therefore suggested long ago that circadian frequency might be achieved in an organism by a process of frequency reduction involving 'beats' set up between such high frequency oscillators.

Beats are the well-known phenomenon in sound that occurs when two pitches of similar frequency add together to produce a secondary, low frequency output by amplitude modulation. Large frequency reductions can be produced in this way, but beat oscillations are much less stable than their primary oscillations and break down rapidly unless these remain absolutely constant – much more constant than the sort of biochemical oscillation shown in Fig. 7–3. Since one of the prime

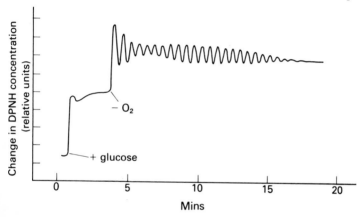

Fig. 7–3 High frequency biochemical oscillations. The oscillation is in the level of intracellular DPNH measured spectrophotometrically in a suspension of yeast cells. The DPNH concentration increases sharply with the addition of glucose to the suspension and then goes spontanesouly into a damped oscillation of about two cycles per min when the O_2 is used up. Although such oscillations in cell suspensions damp out rapidly as shown here, cell-free suspensions fed continuously with substrate (the sugar trehalose) will oscillate at c. 1 cycle/5 min for hundreds of cycles. (Redrawn from PYE, E. K. (1971). *Biochronometry*, ed. M. Menaker. National Academy of Sciences, Washington, p. 623.)

characteristics of circadian and other clock-timed rhythms is their stability in the face of changes in temperature or phase-shifting stimuli (§ 2.6, 2.8), it is unlikely that beat phenomena play a significant role in their clock mechanisms. This does not rule out the possibility of other forms of frequency reduction, however.

One alternative, and apparently inherently more stable means for frequency reduction was suggested by T. Pavlidis in 1969. His theory assumes the existence of inhibitory coupling between high frequency oscillators and proposes that they interact, not to produce beats, but instead to form a kind of single, low frequency multi-oscillator. He explored this model by computer simulation involving populations of up to 30 high frequency oscillators, and found that strong negative coupling between them resulted in sufficient frequency reduction to bring the population down from the sort of frequency shown in Fig. 7–3 to within the circadian range. The strength of this mathematical model is that it is dynamic and, unlike beat phenomena, therefore sufficiently stable for its output to be little affected by small changes in its components. It thus provides a plausible theoretical basis for the origin of circadian (or tidal) rhythmicity within cells. On the other hand, the biochemistry of the necessary interactions between the assumed intracellular biochemical oscillations is quite unknown. There is, however, one example of a multicellular system that may be a working model of it. This concerns the circadian rhythm of electrical activity emitted by the simple eyes of the sea hare, *Aplysia*.

The eyes can be removed from *Aplysia* and kept alive for several days *in vitro*. Such an eye, cultured in constant darkness, produces a clear circadian rhythm in the rate of compound action potentials discharged down its cut optic nerve (Fig. 7–4c). Its retina can then be cut away without significantly upsetting the rhythm until the numbers of the interneurons which produce the action potentials are reduced from their original *c*. 1000 to *c*. 100–200; any further reduction then leads to the rhythm's breakdown (Fig. 7–4d).

Two explanations of this observation are possible. Either, there are a few pacemaker neurons near the base of the optic nerve, each a kind of electrical clock, driving the interneurons to spike in a circadian rhythm. Or, there is a minimum population of *c*. 200 interneurons which, though the individual cells are high frequency oscillators, interact Pavlidis-style to produce a net low frequency, circadian output. If this latter explanation is correct the *Aplysia* eye will provide not only a multicellular vindication of Pavlidis' hypothesis, but also the first working model of a circadian clock.

7.4.3   Membrane permeability cycles

We are still a long way from observing any clock mechanism directly, but two aspects of them are indirectly accessible: (1) their spontaneous, free-running period, and (2) their phase-shifting responses to entraining stimuli. These two are as much direct measures of the underlying oscillator as is the information a doctor gains about his patient's heart from listening to its beat with a stethoscope. Only when an experimental manipulation has affected one of these two parameters can one be sure that the experiment has touched the clock's oscillator mechanism rather than just some aspect of its hands (see § 2.8).

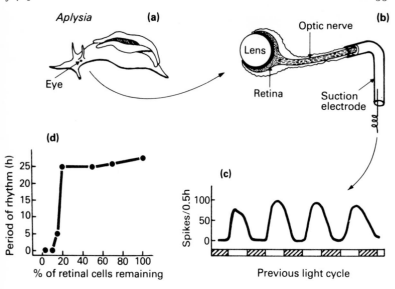

Fig. 7–4   Circadian rhythm of electrical activity in the eye of the sea hare, *Aplysia*. (a) Whole animal showing position of left eye. (b) Dissected eye showing recording method. The eye consists of a spherical lens largely surrounded by a retina of about 4000 photoreceptor cells (black area) and 1000 interneurones (stippled) leading into the optic nerve; the whole structure is covered in a connective tissue sheath. Regular bursts of compound action potentials (spikes) generated by the interneurones can be recorded for days in an isolated eye maintained in tissue culture. (c) First 4 days of the circadian rhythm of spike production by one isolated eye kept in constant darkness; abscissa indicates light cycle experienced by the animal until the eye was removed. (d) Effect of removing different amounts of the retina on the period of the isolated eye's rhythm in constant darkness. The rhythm remains circadian until there are only about 200 (20%) of the interneurones left connected to the optic nerve; any further reduction of these cells causes the rhythm to collapse (§ 7.4.2). (Compiled from JACKLET, J. and GERONIMO, J. (1971). *Science*, **174**, 299; and JACKLET, J. (1974). *J. comp. Physiol.*, **90**, 33.)

Very few chemical agents have these effects, but heavy water (D_2O), alcohol (MeOH, EtOH) and lithium ions all slow down the rhythms they have been tested on (i.e. they lengthen the free-running period), and potassium ions and valinomycin cause phase-shifts. Heavy water, alcohol and lithium include among their biological actions the ability to change membrane permeabilities. By this means, for example, heavy water slows down the membrane-based oscillation involved in the high frequency electric discharge rhythm (900 Hz) of the electric eel, *Stenarchus*. This effect has therefore been invoked to explain the slowing down of circadian rhythms that heavy water, alcohol and lithium cause in plants (beans),

Protista (*Euglena*), arthropods (beach isopods), birds (waxbills), and mammals (rats).

If membranes are involved, then they will presumably be so via their regulation of ion fluxes and gradients between cells or cellular compartments, and will involve cyclical changes in ion movements. Artificially changing crucial ion concentrations in such systems should therefore upset the ionic balance on the two sides of the membrane at some inappropriate moment and thereby create a different phase in the cyclical passage of ions across it. If the clock is based on such ionic movements, this sort of manipulation should cause phase-shifts in it that will be detectable in its 'hands' – the overt rhythms it controls. It is therefore intriguing that pulses of high potassium concentration cause phase-shifts both in the leaf movement rhythm of the bean and in the action potential rhythm of isolated *Aplysia* eyes. More impressively, treatment with valinomycin (an antibiotic that changes membrane permeabilities and acts as a specific carrier for monovalent cations, particularly for potassium) causes both the bean leaf rhythm and the luminescence rhythm of *Gonyaulax* (Fig. 4–1) to phase-shift in parallel with the way they phase-shift to pulses of light or high potassium.

This kind of evidence rather strongly implies the participation of membranes in circadian clock mechanisms. Exactly how is not yet clear, but one fairly general model has been put forward by D. Njus, F. M. Sulzman and J. W. Hastings (1974, *Nature, Lond.*, **248**, 116). This is based on the fluid mozaic theory of membranes and assumes ion transport channels across the lipids of the membrane. It is proposed that these channels operate in a concentration-dependent feedback system to create a slow cyclical change in the ion concentrations on the two sides of the membrane. All the components of the model are known to exist. For example, photosensitive ion channels potentially suitable for phase-shifting and entrainment to light signals are known in plants (e.g. phytochrome, § 6.4) and in animals (in retinal cells). Similarly, the known ability of membranes to maintain a stable fluidity at different temperatures provides for a potential temperature compensation system.

The strength of the hypothesis thus rests on three grounds. (1) There is much circumstantial evidence for ionic phenomena in circadian timing. (2) The model accounts for temperature compensation and entrainment by reasonably well-known processes. (3) The kind of phenomena proposed operate on a time scale many times slower than those involved in any known biochemical rhythms (cf. Fig. 7–3). At the moment, membrane-based mechanisms therefore seem the most plausible basis for oscillatory clocks of the circadian or circa-tidal kind, though other biochemical components, such as protein synthesis (§ 7.4.1), may well be involved too.

Whether the membrane processes invoked actually operate on the 24-h or 12.4-h time scale has yet to be determined, but even if they do there is no reason to suppose that all biological clocks will run on the same mechanism. The fact that they have similar formal characteristics may simply reflect convergent evolution. Man-made mechanical and electrical clocks, after all, both do the same job by quite different means.

7.5   The overall temporal organization of organisms

It is important to bear in mind that one is dealing with a multi-level phenomenon when considering how the physiology of any organism is timetabled. Broadly, clock-timed biological processes fall into the following strata of physiological organization.

(1) *Cellular* – rhythms in protein synthesis, mitosis, photosynthesis, etc. (§ 4.2).
(2) *Tissue* – more general rhythms in metabolism, hormone secretion, pharmacological sensitivity, etc. (§§ 4.3, 4.4.1).
(3) *Behavioural* – daily-repeated on-going cycles of behaviour such as locomotor activity, drinking, responsiveness to stimuli, navigation, etc. (§§ 4.4.2, 5.2).
(4) *Developmental* – gated, once-in-a-lifetime morphogenetic events such as birth, moulting and death (§ 4.4.3).
(5) *Photoperiodic* – the annual regulation of physiology and development in relation to daylength (§ 6.1) often incorporating dormancy timing (§ 6.5).

In addition, there may be processes that involve circannual and circalunar rhythmicity (§§ 2.4, 2.5).

The coexistence of these different levels of temporal organization within one individual organism raises the question of whether a single 'master clock' should be looked for, or some much more complex system. Unquestionably, the individual cells of multicellular animals and plants are capable of showing independent circadian rhythmicity (§ 4.2). Moreover, the most primitive eukaryotic unicellular organisms studied also have circadian rhythms (§ 4.1). This implies both a phylogenetically primitive origin to rhythms and the probability that every cell in a multicellular organism contains its own clock. Yet certain localized operations, in higher animals at least, can stop all overt circadian rhythmicity: removing the silkmoth's brain, cutting the cockroach's optic tract, or cauterizing the rat's supra-chiasmatic nuclei, for example, do just that (§ 7.3).

How is this paradox of universal cellular rhythmicity but specific tissue control of timing to be resolved? The answer probably lies in a hierarchical organization of the organism's clocks (Fig. 7–5). At the lowest level, the cells retain their primitive time-keeping ability but make little use of it except perhaps for basic metabolic functions, keeping in phase with their neighbours by direct intercellular chemical interactions. At the

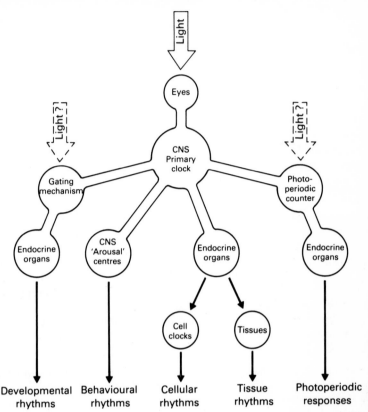

Fig. 7–5 The hierarchy of clocks in an animal. The linked spheres represent direct nervous connections, the black arrows hormonal or other control systems. The central circadian control of behaviour (see p. 33) is envisaged as being effectively 'primary' in all higher animals, acting to drive or entrain their lower-order clocks while being itself entrained via light perceived by the eyes. In some insects the clock gating development acts as its own photoreceptor (without reference to the eyes), and so too may the animal's photoperiodic hourglass (see p. 43) where one exists – hence the dotted light arrows (see below for further explanation).

tissue level, groups of cells whose clock processes have become more highly developed are gathered together and act as driving oscillators, especially in endocrine organs such as the hamster adrenal gland, bird pineal, or silkmoth brain (§ 7.3). At the behavioural level, such concentrations of clock-specializing cells occur in the CNS and act to time behaviour. At the developmental level, the rhythmic gating of growth is probably controlled in some cases by the same primary CNS clocks (e.g. in *Drosophila*, emergence is apparently gated by the clock that times locomotor activity), but in others, parallel clock systems appear to be involved (e.g. in silkmoths, different parts of the brain time emergence and flight activity).

Because so much of the physiology of an animal is heavily influenced by, or actually subserves behaviour, it is reasonable to view the CNS clock or clocks that regulate behaviour as being effectively *the* primary driving oscillator (e.g. the mammalian supra-chiasmatic nuclei – p. 50; or the insect optic lobes – p. 49), and the endocrine clocks (e.g. mammalian adrenals or bird pineal – p. 51) as being effectively secondary.

The arrangement of this hierarchy in plants must be much simpler. A large proportion of their tissues is photosensitive and fully exposed to the light, so it is presumably directly entrained at the cellular level. Their root tissue, however, is also endogenously rhythmic (p. 30), so some kind of hierarchy may well exist. Since auxin levels vary rhythmically across the day, and since auxin treatment affects leaf movements, there may prove to be a form of hormonal coupling between the green tissues and those not exposed to light.

Where photoperiodic timing is based on a circadian rhythm, as it is in soy-beans or quail (p. 45), it presumably involves coupling from the primary or secondary oscillators. Where it involves an hourglass, non-rhythmic mechanism, as it does in the vetch aphid (p. 44), a different set of cells specialized as a clock must exist and operate independently of the circadian hierarchy.

Circannual and circa-lunar timing may likewise be independent systems, or, perhaps more plausibly, may be coupled to some circadian clock and consist of a system that 'adds up' approximately 365, 29, or 15 cycles by accumulating a daily aliquot of some long-lived biochemical. Something like this does occur in several photoperiodic clocks, which have to 'add up' several cycles of inductive night length before development will switch (§ 6.3, last para.), so that this is a reasonable explanation of the mysteries of long-cycle clock mechanisms emphasized in § 3.1. Possibly dormancy-duration timers (§ 6.5) work similarly; since neither their nature nor their anatomical location is known they, too, are omitted from Fig. 7–5.

Overall, therefore, animals and to a lesser extent plants must be considered as hierarchical complexes of clocks. The idea of *a master clock* controlling all their physiological processes is clearly much too simple. On the other hand, in higher animals, the central nervous timing of behaviour may be so all-pervasive in its effects that it appears to act as one. That being so, the system may be visualized as outlined in Fig. 7–5.

Key to Terminology

amplitude, p. 5
autophasing, p. 22
azimuth, p. 37
circadian, p. 11
critical photoperiod, p. 42
DD (cf. LL), p. 9
diapause, p. 46
diel, p. 11
diurnal, p. 11
dormancy, p. 46
endogenous, p. 21
entrainment, p. 16
exogenous, p. 21

free-running, p. 11
frequency, p. 6
gating, p. 35
hourglass, p. 43
LD, p. 9
LL (cf. DD), p. 9
neap tides, p. 12
nocturnal, p. 11
oscillator, p. 7
period, p. 5
phase, p. 6
phase response curve, p. 20
phase-shift, pp. 6 and 16

photoperiodism, p. 41
phytochrome, p. 45
Q_{10}, p. 15
rhythm, p. 7
sine wave, p. 7
spring tides, p. 12
sun-compass, p. 37
synodic, p. 12
temperature compensation, p. 15
zeitgeber, p. 16
zugunruhe, p. 38

Further Reading

BÜNNING, E. (1967). *The Physiological Clock*, 2nd edition. Springer-Verlag, New York.

CONROY, R. T. W. L. and MILLS, J. N. (1970). *Human Circadian Rhythms*. Churchill, London.

KENDRICK, R. E. and FRANKLAND, B. (1976). *Phytochrome and Plant Growth*. Studies in Biology no. 68, Edward Arnold, London.

LEES, A. D. (1955). *The Physiology of Diapause in Arthropods*. Cambridge University Press, Cambridge.

LOFTS, B. (1970). *Animal Photoperiodism*. Studies in Biology no. 25, Edward Arnold, London.

MATHEWS, G. V. T. (1968). *Bird Navigation*, 2nd edition. Cambridge University Press, Cambridge.

OSWALD, I. (1966). *Sleep*. Pelican, London.

PALMER, J. D. (1974). *Biological Clocks in Marine Organisms*. John Wiley, New York and London.

PALMER, J. D., BROWN, F. A., JR., and EDMUNDS, L. N., JR. (1976). *An Introduction to Biological Rhythms*. Academic Press, New York and London.

'PUDOC' (1972). *Circadian Rhythmicity*. Proceedings of the International Symposium on Circadian Rhythmicity, Wageningen 1971. Pudoc, Wageningen, Holland.

SAUNDERS, D. S. (1977). *An Introduction to Biological Rhythms*. Blackie, Glasgow and London.

SWEENEY, B. M. (1969). *Rhythmic Phenomena in Plants*. Academic Press, London and New York.

VINCE-PRUE, D. (1975). *Photoperiodism in Plants*. McGraw-Hill, London and New York.